matrix algebra:

a programmed introduction

RICHARD C. DORF
dean, college of engineering
ohio university

john wiley & sons, inc.

new york, london, sydney, toronto

10 9 8 7 6 5 4 3 2 1

SBN 471 21908 8 (cloth) SBN 471 21909 6 (paper)
Printed in the United States of America

preface

Matrix algebra has now become an integral part of the mathematical background necessary for engineers and scientists. Furthermore, knowledge of the fundamental methods of matrix algebra is necessary for social scientists, economists, and students of education and business.

The primary purpose of this book is to enable the reader to develop a skill in utilizing the algebra of matrices. Thus, we are primarily concerned with providing a working knowledge of matrix algebra and its application to important problems that arise in engineering, economics, sociology, and computer science, to name just a few.

This book will serve as a supplemental aid for students who find in various courses that a knowledge of the algebra of matrices is presumed or would be helpful. Although it is often inappropriate at that point to undertake a course in matrix theory, the use of this text will enable the student to develop the necessary skill with matrices.

This textbook is written in a programmed format that should aid in the achievement of its primary goal. This format provides each reader with an individual path through the subject material and is particularly useful for self-study. In addition, the student will proceed at the pace he finds most comfortable. Thus, the book may be used by students with a wide range of abilities, backgrounds, and reading speeds. Hopefully the author will act as an individual tutor by leading the student from the simple to the more complex material, required active participation by the student, and by providing opportunities for testing the knowledge gained.

This book should be particularly useful for students who are studying or taking a course in any of the following subjects:

Electrical circuit analysis
Analysis of linear systems
Mechanical vibrations
Control systems
Structural mechanics and analysis
Numerical methods and computer science
State variables and time-domain methods
Economics, particularly econometrics
Mathematics for social scientists
Mathematics with business applications

OBJECTIVES

Upon completion of this book, the reader should be able to utilize the algebra of matrices in solving problems in various disciplines in which he has an adequate knowledge. Specifically, he should be able to

1. Identify a matrix and its order
2. Add two or more matrices
3. Multiply two or more matrices
4. Evaluate a minor, cofactor, and the determinant of a matrix
5. Identify and discriminate among the identity, null, symmetric, and diagonal matrix
6. Obtain the transpose of a matrix
7. Obtain a linear transformation of a matrix
8. Evaluate the adjoint matrix and then the inverse of a matrix
9. Obtain the matrix representation of a set of linear algebraic equations and solve the set of equations for unknown variables by utilizing matrix inversion
10. Evaluate the rank of a matrix
11. Obtain the characteristic roots of a matrix and the corresponding characteristic vectors
12. Diagonalize a matrix using a matrix transformation
13. Obtain the exponential series for a matrix
14. Differentiate a matrix

PREREQUISITES

It is expected that the reader of this book possesses a knowledge of high-school algebra. The book should be of use to those with only the rudiments of algebra as well as to engineering students who have completed two years of college calculus and differential equations. The minimum mathematical knowledge required is that of algebra and the fundamental algebraic concepts of addition, multiplication, and the solution of a linear algebraic equation for an unknown variable. It is also necessary for the reader to know how to factor quadratic equations and how to solve them in one or two unknowns. Furthermore, it will be helpful if he can factor higher order polynomials.

Use the following questions to test your knowledge of the prerequisite material. The correct answers follow the test questions.

1. Complete the following operation.

$$(a + b)(a - b) = \underline{\hspace{6cm}}$$

2. Factor the following equation.

$$x^2 - 7x + 6 = \underline{(\quad)(\quad)\qquad}$$

3. Factor the following equation.

$$x^2 + 2xy - 8y^2 = \underline{\qquad\qquad}$$

4. Evaluate x^{-3} when $x = 2$. ______________

5. Solve the following linear algebraic equation for x.

$x + 7 = 5 \qquad x =$ ______________

6. The number $3 + 6j$ where $j = \sqrt{-1}$ is called a ______________ number.

7. Solve $3x^2 - 5x + 1 = 0$ for x. $x =$ ______________

Answers

1. $a^2 - b^2$
2. $(x - 1)(x - 6)$
3. $(x + 4y)(x - 2y)$
4. $x^{-3} = 2^{-3} = \frac{1}{2^3} = \frac{1}{8}$
5. $x = -2$
6. complex
7. $x = \frac{5}{6} \pm \frac{\sqrt{13}}{6}$

If you have completed four or more questions correctly you are ready to proceed to the subject considered in this text. If you completed fewer than half the questions correctly, you probably should review your high-school algebra first.

HOW TO USE THIS BOOK

This book is quite different from most textbooks you have used. It is in a programmed format and requires a different reading procedure. A programmed text provides an opportunity for dialogue with the author. You may consider the author of this book as your tutor. Furthermore, you are required to work through this book and answer questions at each stage. Therefore, your developing skill will be tested continually, and you will be able to proceed at a rate commensurate with your ability at each stage of the program. It is necessary that the material of each section be thoroughly understood before you continue.

The procedure for using this book is

1. Read the information and examples provided in each frame.
2. Complete each problem at the end of each set of frames.
3. Use the answer shield provided with this program to cover the correct answer in the gray area following each frame. After you have completed your answer, slide the shield down to compare your response to the printed answer.
4. Read the explanation of the answer if your answer was incorrect or you are not confident of the process you used to obtain the answer, and then proceed to the next frame.
5. On a few pages you are asked to choose between several answers which lead to different following pages for the correct answer and explanation. Just follow the instructions on these pages and you will not get lost.
6. Do only a portion of the chapter at each sitting. Suitable places to stop within a chapter will be indicated. Remember, there is a large amount of material contained in these pages and ten to twenty separate sittings are required for the completion of the text. Take your time, it will be worth it!

Well, we are both ready to begin, so let's turn to the first chapter.

contents

Chapter 1	*introduction to matrices*	1
Chapter 2	*types of matrices*	21
Chapter 3	*matrix operations*	37
	Addition	41
	Subtraction	45
	Matrix Multiplication	48
Chapter 4	*linear equations and determinants*	62
	Determinants	68
	Cramer's Rule	82
	Elimination Method	86
Chapter 5	*the rank, trace, and adjoint of a matrix*	96
	The Trace of a Matrix	107
	Cofactor Matrix	108
Chapter 6	*the inverse of a matrix*	121
	The Orthogonal Matrix	136
Chapter 7	*the characteristic equation of a matrix*	148
	Linear Dependence	165

Chapter 8 *matrix transformations and functions of a matrix* 190

The Derivative of a Matrix 218

answers to exercises 236

index 259

1 introduction to matrices

1-1

Many concepts can be described as *quantitative* because they can be described by using numbers. In this text we are concerned with the characteristics of nature and man's world that can be described quantitatively. For example, we may describe a person's age quantitatively since age is stated as a specific number of years.

Which of the following cannot be described quantitatively?
(a) length of an automobile
(b) speed of a train
(c) personality

____

(c) Personality is, of course, the correct answer. Length of an automobile and speed of a train can be described in terms of feet or miles per hour, respectively. There is no meaningful way, at present, for assigning a quantitative value or number to a psychological term such as personality. As you noted, the length, or speed of an object is a quantitative characteristic of the object.

1-2

A quantity of specific interest in mathematics is one called a *scalar quantity.* A scalar quantity is one that can be identified by a single number. An example of a scalar quantity is the speed of a vehicle.

Are any of the following not scalar quantities?
(a) temperature of an object
(b) population of a town
(c) the length of a stick

____

I hope you answered no. All those mentioned are scalar quantities since they may be described by a single number. We may find that the temperature of an object is 75 degrees Fahrenheit, the population of Athens, Ohio is 20,000, and the length of a stick is three feet.

1-3

Well then, what is an example of a nonscalar quantity? Such an example is the description of the size of this page. In order to quantitatively describe the size of this page we need two numbers, the length and the width. Because we need a pair of numbers, we use a nonscalar quantity. Therefore, we need the width of 8 1/2 inches and length of 11 inches to represent the size of the page as the pair of number 8 1/2 and 11. We can write this pair of numbers in the following manner:

$$(8\ 1/2, 11)$$

Thus, we will choose to represent a pair of associated numbers within parentheses set off from each other by a comma. If we choose to represent a set of more than two associated numbers we will again use parentheses to incorporate the set of numbers.

Give a quantitative description for the three associated quantitatives of a rectangular box: the length, l; the width, w; and the depth, d.

____

You are correct if you described the length, width, and depth of a rectangular box as follows:

$$(l, w, d)$$

We have included the set of three quantities within the parentheses and set them apart by commas. Thus, we have written the quantitative description of the box by writing the prescribed order, writing the symbols l, w, and d, and by placing them within parenthese. If we know that the box was 8 inches long, 4 inches wide and 3 inches deep we could write the quantitative description as

$$(8, 4, 3)$$

1-4

Now write a similar quantitative description for a rectangular box of length l, width w, and depth d, which contains n objects.

You are correct if you obtained the representation

$$(l,\ w,\ d,\ n)$$

to describe the four quantities that represent the length, width, and depth of the box and the number of objects within.

1-5

The scalar quantities contained within the parentheses are called the *elements* of the *array*. That is, we have agreed to include within the parentheses the elements l, w, d, and n. Furthermore, we note that these elements are written in a horizontal row and in an agreed upon order with length as the first element and the number of objects within the box as the last element of the row.

Which of the following is a row array of three elements?

(a) $(h,\ i,\ j)$

(b) $\begin{pmatrix} h, & i \\ j, & k \end{pmatrix}$

(c) $\begin{pmatrix} h \\ i \\ j \end{pmatrix}$

____

You are correct if you selected (h, i, j) as a row array of three elements. The matrix in (b) contains four elements in two rows. The matrix in (c) contains three elements but is a column array, not a row.

1-6

Now let us consider two rectangular boxes labelled box 1 and box 2. The row array representing the length, width and depth of the first box may be written as

$$(l_1,\ w_1,\ d_1)$$

where l with the subscript 1 refers to the length of the first box. Similarly, the length, width, and depth of the second box is represented by the row array as

$$(l_2,\ w_2,\ d_2)$$

Now we may combine the two rows within one pair parentheses with the first row for the first box and the second row for the second box as follows:

$$\begin{pmatrix} l_1, & w_1, & d_1 \\ l_2, & w_2, & d_2 \end{pmatrix}$$

Thus, we have obtained an array with two rows and six elements.

Write an array for these two rectangular boxes, which represents the width and number of objects within each box.

$$\begin{pmatrix} w_1, & n_1 \\ w_2, & n_2 \end{pmatrix}$$

If you obtained this correct answer proceed to Frame 1-7. If you obtained an incorrect answer or want a further explanation, read the following paragraph.

The foregoing answer contains two rows and four elements. We note that the first row contains the width of the first box, w_1, and the number of objects within the box, n_1. Similarly, the second row contains w_2 and n_2. If you have any questions about this answer, reread Frame 1-6 and then proceed to 1-7.

1-7

Now let us describe the complete set of attributes for the two boxes. The attributes of the two boxes, that is, length, width, depth, and number of objects within the box, are represented by the eight-element array as

$$\begin{pmatrix} l_1, & w_1, & d_1, & n_1 \\ l_2, & w_2, & d_2, & n_2 \end{pmatrix}$$

If we wished to describe one additional attribute for each box, we then would have ____________________ elements within the parentheses.

Ten elements are required within the parentheses. Each box would then have five attributes and therefore we would need ten elements to represent the two boxes.

1-8

If we add the attribute of the thickness of the walls of a box, t, the representation of the two boxes would be

$$\begin{pmatrix} l_1, & w_1, & d_1, & n_1, & t_1 \\ l_2, & w_2, & d_2, & n_2, & t_2 \end{pmatrix}$$

Thus, the number of attributes of each box can be increased by adding an additional element at the end of each row.

Extending our present understanding, what will be the representation in array form for three boxes, when we are concerned with the four attributes of length, width, depth, and wall thickness of each box?

You are correct if you obtained the matrix.

$$\begin{pmatrix} l_1, & w_1, & d_1, & t_1 \\ l_2, & w_2, & d_2, & t_2 \\ l_3, & w_3, & d_3, & t_3 \end{pmatrix}$$

to represent the four attributes of the three boxes.

1-9

Now, at this point it is clear that we could drop the use of commas to set off each element, and write the array as

$$\begin{pmatrix} l_1 & w_1 & d_1 & t_1 \\ l_2 & w_2 & d_2 & t_2 \\ l_3 & w_3 & d_3 & t_3 \end{pmatrix}$$

The element w_2, for example, is still considered to be set off from the surrounding elements. The deletion of the commas is standard practice in textbooks and journals.

A more general representation of this array can be obtained by noting that the first element of each row represents the first attribute of the box. Therefore, if we are interested only in the first attribute of three boxes we would obtain the vertical array

$$\begin{pmatrix} l_1 \\ l_2 \\ l_3 \end{pmatrix}$$

This array forms a *column* of three elements.

Write an array for the two attributes length and width for three boxes.

You are correct if you wrote

$$\begin{pmatrix} l_1 & w_1 \\ l_2 & w_2 \\ l_3 & w_3 \end{pmatrix}$$

This array possesses six elements and has three rows and two columns. There is a row for each box and a column for each attribute.

1-10

Now let us develop a scheme for labelling the elements of an array more generally. Each element of the array will be labelled as a lower case letter with a double subscript. In general, a is used as the letter of the array. The elements of the first row are labelled a_{1j}. The subscript one refers to the first row; the second subscript, j, refers to the column number. The, for example, the first box with four attributes, l, w, d, and n, is represented by the row array

$$a_{11} \quad a_{12} \quad a_{13} \quad a_{14}$$

which is equivalent to

$$(l_1 \quad w_1 \quad d_1 \quad n_1)$$

Use the subscripted variable a to write the array that represents two boxes with four attributes.

$$\begin{pmatrix} a_{11} & a_{12} & a_{13} & a_{14} \\ a_{21} & a_{22} & a_{23} & a_{24} \end{pmatrix}$$

is the correct answer. Note that each row has four elements. If you had difficulty in obtaining this answer, reread this frame, and then attempt to answer the question again. You should have noted that the jth column contains the jth attribute for each box.

1-11

A single element that lies at the intersection of the ith row and the jth column is denoted as a_{ij}. Then, for example, a_{12} lies at the intersection of the first row and the second column of the array.

In order confirm your understanding of the representation in an array format, complete the following question.

Each array contains three rows and three columns, with the subscripts on all but one element correctly indicated. Circle the incorrect element of each array.

(a) $$\begin{pmatrix} a_{11} & a_{12} & a_{13} \\ a_{12} & a_{22} & a_{23} \\ a_{31} & a_{32} & a_{33} \end{pmatrix}$$

(b) $$\begin{pmatrix} a_{11} & a_{12} & a_{13} \\ a_{21} & a_{22} & a_{23} \\ a_{31} & a_{23} & a_{33} \end{pmatrix}$$

(c) $$\begin{pmatrix} a_{11} & a_{12} & a_{13} \\ a_{21} & a_{22} & a_{23} \\ a_{31} & a_{32} & a_{31} \end{pmatrix}$$

The correct array with three rows and three columns is

$$\begin{pmatrix} a_{11} & a_{12} & a_{13} \\ \boxed{a_{21}} & a_{22} & a_{23} \\ a_{31} & \boxed{a_{32}} & \boxed{a_{33}} \end{pmatrix}$$

Position of error in (a) — Position of error in (b) — Position of error in (c)

1-12

The rectangular array of numbers generally will possess m rows and n columns and will be written as

$$\begin{pmatrix} a_{11} & a_{12} & a_{13} & \cdots & a_{1n} \\ a_{21} & a_{22} & a_{23} & \cdots & a_{2n} \\ \cdot & \cdot & \cdot & \cdot & \cdot \\ \cdot & \cdot & \cdot & \cdot & \cdot \\ \cdot & \cdot & \cdot & \cdot & \cdot \\ a_{m1} & a_{m2} & a_{m3} & \cdots & a_{mn} \end{pmatrix}$$

The dots within the array indicate that we have not explicitly written all the elements.

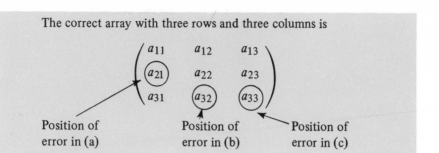

Write an array that possesses four rows and two columns.

The array with four rows and two columns is

$$\begin{pmatrix} a_{11} & a_{12} \\ a_{21} & a_{22} \\ a_{31} & a_{32} \\ a_{41} & a_{42} \end{pmatrix}$$

1-13

Well, we are now in a position to be able to readily define a matrix. A *matrix* is a rectangular array of numbers (scalar quantities) that are represented as follows:

$$\begin{pmatrix} a_{11} & a_{12} & \cdots & a_{1n} \\ a_{21} & a_{22} & \cdots & a_{2n} \\ \cdot & \cdot & \cdot & \cdot \\ \cdot & \cdot & \cdot & \cdot \\ \cdot & \cdot & \cdot & \cdot \\ a_{m1} & a_{m2} & \cdots & a_{mn} \end{pmatrix}$$

So, you see, we have been looking at matrices (plural of matrix) for several pages now.

It is standard to use brackets instead of large parentheses around the array of numbers, so that the matrix is usually written as

$$\begin{bmatrix} a_{11} & a_{12} & \cdots & a_{1n} \\ a_{21} & a_{22} & \cdots & a_{2n} \\ \cdot & \cdot & \cdot & \cdot \\ \cdot & \cdot & \cdot & \cdot \\ \cdot & \cdot & \cdot & \cdot \\ a_{m1} & a_{m2} & \cdots & a_{mn} \end{bmatrix}$$

The numbers a_{11}, a_{12}, ... are called the *elements* of the matrix. The first subscript of each element denotes the row and the second subscript denotes the column of the matrix in which the element lies.

The foregoing matrix possesses a total of ______ rows and __________ columns.

If you said that the matrix

$$\begin{bmatrix} a_{11} & a_{12} & \cdots & a_{1n} \\ a_{21} & a_{22} & \cdots & a_{2n} \\ \cdot & \cdot & & \cdot \\ \cdot & \cdot & & \cdot \\ \cdot & \cdot & & \cdot \\ a_{m1} & a_{m2} & \cdots & a_{mn} \end{bmatrix}$$

possesses a total of m rows and n columns, you are correct.

1-14

Write a matrix with three rows and four columns. How many elements does this matrix possess? ____________________

A matrix with three rows and four columns is written as

$$\begin{bmatrix} a_{11} & a_{12} & a_{13} & a_{14} \\ a_{21} & a_{22} & a_{23} & a_{24} \\ a_{31} & a_{32} & a_{33} & a_{34} \end{bmatrix}$$

This matrix possesses twelve elements. You can readily show that this matrix has 3 X 4 or twelve elements.

1-15

How many elements are there in a general matrix with m rows and n columns? ____________________

There are $m \times n$ or m times n elements in a matrix with m rows and n columns. The number of elements in a matrix is finite when, as is the usual case, m and n are finite numbers.

1-16

The matrix of m rows and n columns is said to be of *order* m by n, often written as $m \times n$. Giving the order of a matrix is a way to define the size of the matrix.

(a) $\begin{bmatrix} a_{11} & a_{12} & a_{13} \\ a_{21} & a_{22} & a_{23} \end{bmatrix}$ (b) $\begin{bmatrix} a_{11} & a_{12} \\ a_{21} & a_{22} \end{bmatrix}$

Which matrix is at order 3 X 2: (a); (b); neither? Choose one answer.

____

You are correct if you said that neither matrix is of order 3×2. Matrix (a) is of order 2×3 since there are two rows and three columns. Matrix (b) is of order 2×2.

1-17

The matrix with an equal number of rows and columns has a special name apropos of its form. A matrix with n rows and n columns is called a *square matrix of order n.* Note that the square matrix is a special case of the rectangular matrix defined in Frame 1-13.

Write a square matrix of order 3.

Yes, a square matrix of order 3 is

$$\begin{bmatrix} a_{11} & a_{12} & a_{13} \\ a_{21} & a_{22} & a_{23} \\ a_{31} & a_{32} & a_{33} \end{bmatrix}$$

1-18

We have been using the general elements a_{ij} for the last several pages. However, we should not forget that these elements will be numbers for the specific matrices that arise in different problems. Thus, a specific square matrix of order 2 might be

$$\begin{bmatrix} 2 & 1.1 \\ -3 & 0 \end{bmatrix}$$

Therefore, in this case we have $a_{11} = 2, a_{12} = 1.1, a_{21} = -3$, and $a_{22} = 0$. The elements may attain any numerical value. In many cases in engineering the element values may even be complex numbers. Remember that a matrix is simply a rectangular array of numbers.

Write a square matrix of order 3 in which every alternate element along a row or column is a 0 or a 1 and $a_{11} = 1$. That is, the elements alternate from 1 to 0 and back to 1.

$$\begin{bmatrix} 1 & 0 & 1 \\ 0 & 1 & 0 \\ 1 & 0 & 1 \end{bmatrix}$$

is the answer for a square matrix of order 3 where the elements alternate from 1 to 0 and then to 1.

1-19

We will be able to use a compact notation for the matrices by assigning a boldface capital letter to each matrix*. Therefore we could distinguish two matrices **A** and **B** as follows:

$$\mathbf{A} = \begin{bmatrix} 3 & 2 & 1 & 5 \\ 4 & 6 & -2 & 1 \\ 1.2 & 0 & 1 & 8 \\ 6 & 2 & 0 & 0 \end{bmatrix} \qquad \mathbf{B} = \begin{bmatrix} 4 & -1 \\ 8 & 2 \\ 6 & -2 \end{bmatrix}$$

*The student can use a wavy line under the letter to indicate boldface: for example, $\utilde{A}$

Which is a square matrix?

What is its order? ______________________

A is a square matrix of order 4. (If you said **A** is of order 16 read page 16 before proceeding.)

Since **B** has three rows and two columns, it couldn't be a square matrix, which always has an equal number of rows and columns. If you gave **B** as the answer, reread Frames 1-17 and 1-18 before proceeding.

1-20

Now, to be certain we are able to correctly write a matrix that represents a specific array, let us consider one more familar situation.

The weather is measured in three locations in a city. At each location the maximum and minimum temperature and the precipitation are measured. On a specific day, the temperatures and rainfall measured at the first location are 81, 52, and 0.12; at the second location the measured values are 88, 62, and 0.21; and at the third they are 83, 58, and 0.30.

Construct a matrix to represent this set of data in the form we have been using in this section. What is the order of this matrix? Is this a square matrix?

____

Turn to page 17.

You correctly stated that **A** is a square matrix but indicated it was of order 16. Recall a matrix is in general of order $m \times n$, where m is the number of rows and n is the number of columns in **A**. For a square matrix we have an $n \times n$ matrix and we say it is of order n. Therefore matrix **A** in Frame 1-19 is of order 4. If you obtained 16 as the order of **A** you incorrectly multiplied 4 by 4. The statement 4 by 4 (or 4×4) should not imply multiplication but rather a form to represent the order of a matrix and for the square matrix we abbreviate the notation to say "of order 4." Return to page 15.

You are correct if the matrix you obtained was

$$\begin{bmatrix} 81 & 52 & 0.12 \\ 88 & 62 & 0.21 \\ 83 & 58 & 0.30 \end{bmatrix}$$

This is a square matrix of order three. If you were not able to obtain this answer read page 18 and then return to Frame 1-20. Please note that, as agreed in the previous pages, we use the first row to indicate the attributes of the first location. In each row we list the high and low temperature and the precipitation in that order. This matrix is a compact and orderly representation of the weather at the three locations in the city.

Turn to page 19

We agreed to represent the weather of a city in matrix form. Let's use a matrix to represent the weather at two locations in a city where the maximum temperature, minimum temperature, and precipitation are recorded for each location. We will list the data in that order. Then, if at the first location we have 70°, 50°, and 0.6 inches and at the second location we record 65°, 48°, and 0.4 inches, we will construct the matrix as follows:

	Maximum temperature	Minimum temperature	Precipitation
Location 1	70	50	0.6
Location 2	65	48	0.4

(The units of the elements, that is, degrees and inches, are assumed.)

Return to Frame 1-20 and try the problem again.

SUMMARY

You have now completed the first section of this book. You should now be able to

1. Construct a matrix representation of a set of data
2. Identify a matrix
3. Identify the elements of a matrix **A** as a_{ij}
4. Recognize the order of a matrix

In summary, you have learned that

1. A matrix is a rectangular array of numbers that can be represented as

$$\mathbf{A} = \begin{bmatrix} a_{11} & a_{12} & \cdots & a_{1n} \\ a_{21} & a_{22} & \cdots & a_{2n} \\ \cdot & \cdot & \cdot & \cdot \\ \cdot & \cdot & \cdot & \cdot \\ \cdot & \cdot & \cdot & \cdot \\ a_{m1} & a_{m2} & \cdots & a_{mn} \end{bmatrix}$$

2. The scalar quantities contained within the array are called the elements of the matrix.
3. The order or size of a matrix is determined by the number of rows and columns; thus the matrix **A** in item 1 is of order m by n (or $m \times n$)
4. A square matrix has the same number of rows and columns ($m = n$) and is of order n.

To be certain you are completely familiar with these concepts, work the following exercises.

EXERCISES

1. Which of the following is not a matrix? ______________

$$\mathbf{A} = \begin{bmatrix} 3 & 1 & 8 \\ 2 & 1 & 0 \\ 0 & -3 & 1 \end{bmatrix} \qquad \mathbf{B} = \begin{pmatrix} b_{11} & b_{12} \\ b_{21} & b_{22} \\ b_{31} & b_{32} \end{pmatrix} \qquad \mathbf{C} = \begin{vmatrix} c_{11} & c_{12} \\ & c_{22} \end{vmatrix}$$

2. The order of matrix **B** is ______________ .

3. Matrix ______________ above is a square matrix.

4. Element a_{23} of matrix **A** is equal to ______________ .

5. We wish to represent three characteristics of two men. The characteristics are age, height, and weight. The first man, Mr. Martin, is 41 years old, 73 inches tall, and weighs 172 pounds. The second man, Mr. Lloyd, is 33 years old, 68 inches tall, and weighs 158 pounds. Construct a matrix **M** that will represent the data

   M =

   Matrix **M** is of order ______________ .

The answers appear on page 236.

2 types of matrices

In this chapter you will learn the names of commonly encountered matrices and develop an ability to discriminate among them. In other words, you will expand your vocabulary in the language of matrices so that you will be able to recognize a matrix by its name.

2-1

You are already familiar with two kinds of matrices. We have identified the square matrix of order n and the more general rectangular matrix of order $m \times n$. Now it will be useful to assign names to several specific matrices that are members of the general class of $m \times n$ matrices.

A matrix **A** of order one $\times$ n is written as

$$\mathbf{A} = [a_{11} \quad a_{12} \quad a_{13} \quad \dots \quad a_{1n}]$$

This matrix, which contains n elements and is of order $1 \times n$ is called a *row vector*. Clearly, this matrix is comprised of only one row and thus the use of the word row in the name. The word *vector* is used for matrices that have only one row or column. The row vector still fits the definition of a matrix since it is a rectangular array of numbers even if the number of elements in each column is one.

When an array contains only one row or one column it is more usual to use only one subscript on the elements of the matrix, so that the row vector is commonly written as

$$\mathbf{a} = [a_1 \quad a_2 \quad a_3 \quad \dots \quad a_n]$$

Notice that we have used a boldface lower-case letter to denote this row vector. This is common practice.

Write a row matrix **x** that is of order 1 X 4.

$$\mathbf{x} = [x_1 \quad x_2 \quad x_3 \quad x_4]$$

for a row vector of order 1 X 4.

2-2

Similarly, a *column vector* is a matrix of order $m \times 1$ that contains m rows and one column. We will write a column vector as

$$\mathbf{y} = \begin{bmatrix} y_1 \\ y_2 \\ \cdot \\ \cdot \\ \cdot \\ y_m \end{bmatrix}$$

where again we use a boldface lower-case letter to represent the vector. Each element of the column vector has one subscript, which indicates the number of the row.

Examine the matrices below and complete the following sentence.

$$\mathbf{z} = [z_1 \quad z_2 \quad z_3 \quad z_4 \quad z_5]$$

$$\mathbf{w} = \begin{bmatrix} w_1 \\ w_2 \end{bmatrix}$$

Matrix ____________ is a row vector of order ____________, whereas matrix ____________ is a column vector of order ____________.

Matrix **z** is a row vector of order 1 X 5 and matrix **w** is a column vector of order 2 X 1. If you had any trouble obtaining the answer, return to the beginning of this chapter and review these few pages before proceeding.

2-3

We would like to define a matrix that correcponds to zero in the algebra of numbers. A matrix, every element of which is zero, is called a *zero matrix*. Thus a zero matrix of order 3 X 2 is written as

$$\mathbf{0} = \begin{bmatrix} 0 & 0 \\ 0 & 0 \\ 0 & 0 \end{bmatrix}$$

Another name commonly used for the zero matrix is the *null matrix*.

Write a zero matrix of order 2 X 4.

A zero matrix of order 2 X 4 is

$$\mathbf{0} = \begin{bmatrix} 0 & 0 & 0 & 0 \\ 0 & 0 & 0 & 0 \end{bmatrix}$$

2-4

We would also like to define a matrix that corresponds to 1 in the algebra of numbers. However, first let us specify a matrix that possesses elements other than zero only on the diagonal of the matrix. A *square* matrix **D** whose elements $d_{ij} = 0$ when i is not equal to j ($i \neq j$) is called a *diagonal matrix*, and is written as

$$\mathbf{D} = \begin{bmatrix} d_{11} & 0 & 0 & \cdots & 0 \\ 0 & d_{22} & 0 & \cdots & 0 \\ 0 & 0 & d_{33} & \cdots & 0 \\ \vdots & \vdots & \vdots & & \vdots \\ 0 & 0 & 0 & \cdots & d_{nn} \end{bmatrix}$$

Write a diagonal matrix, **I**, of order n, in which all the elements that lie on the diagonal equal 1.

A diagonal matrix, **I**, of order n where all the elements that lie on the diagonal are equal to 1 is written as

$$\mathbf{I} = \begin{bmatrix} 1 & 0 & 0 & \dots & 0 \\ 0 & 1 & 0 & \dots & 0 \\ 0 & 0 & 1 & \dots & 0 \\ \vdots & \vdots & \vdots & & \vdots \\ 0 & 0 & 0 & \dots & 1 \end{bmatrix}$$

2-5

This particular diagonal matrix with the diagonal elements equal to 1 is called the *identity matrix* (called the unit matrix in several texts). The algebraic operations involving the identity matrix **I** are similar to those we are familiar with involving the number 1 in the algebra of numbers.

A matrix

$$\mathbf{H} = \begin{bmatrix} 1 & 0 \\ 0 & 1 \\ 0 & 0 \end{bmatrix}$$

is an identity matrix. True or False? ____________________

False is the correct answer. A matrix **H** of order 3 × 2 is *not* an identity matrix when $h_{ij} = 1$ for $i = j$ and $h_{ij} = 0$ for $i \neq j$. The identity matrix must be a *square matrix* as we noted on page 30 for a diagonal matrix. The elements on the main diagonal are equal to one and the other elements are zero. Therefore, a matrix of order 3 × 2 cannot be an identity matrix. (If the matrix was a square matrix with $h_{ij} = 1$ for $i = j$ and $h_{ij} = 0$ for $i \neq j$ then **H** would be equivalent to **I**, the identity matrix.)

A square matrix **A** whose elements $a_{ij} = 0$ for i greater than j $(i > j)$ is called an *upper triangular matrix* and is written as

$$\mathbf{A} = \begin{bmatrix} a_{11} & a_{12} & a_{13} & \cdots & a_{1n} \\ 0 & a_{22} & a_{23} & \cdots & a_{2n} \\ 0 & 0 & a_{33} & \cdots & a_{3n} \\ \cdot & \cdot & \cdot & & \\ \cdot & \cdot & \cdot & & \\ \cdot & \cdot & \cdot & & \\ 0 & 0 & 0 & \cdots & a_{nn} \end{bmatrix}$$

That is, all the elements below the major diagonal of an upper triangular matrix are zero. The major diagonal of a square matrix starts in the upper left corner of the matrix and runs down to the lower right corner.

Similarly, a square matrix **B** whose elements $b_{ij} = 0$ for $i < j$ is called a *lower triangular matrix* and for a matrix of order 4 would appear as

$$\mathbf{B} = \begin{bmatrix} b_{11} & 0 & 0 & 0 \\ b_{21} & b_{22} & 0 & 0 \\ b_{31} & b_{32} & b_{33} & 0 \\ b_{41} & b_{42} & b_{43} & b_{44} \end{bmatrix}$$

Refer to matrices **C**, **D**, and **E** and complete the following sentence:

Matrix ________ is an upper triangular matrix of order __________,
whereas matrix ______ is a lower triangular matrix of order _________.

$$\mathbf{C} = \begin{bmatrix} 1 & 1 & 1 \\ 0 & 1 & 1 \end{bmatrix} \qquad \mathbf{D} = \begin{bmatrix} 1 & 0 \\ 2 & 3 \end{bmatrix}$$

$$\mathbf{E} = \begin{bmatrix} e_{11} & e_{12} & e_{13} \\ 0 & e_{22} & e_{23} \\ 0 & 0 & e_{33} \end{bmatrix}$$

Matrix **E** is an upper triangular matrix of order 3; matrix **D** is a lower triangular matrix of order 2. Clearly, the matrix **C** does not qualify as a triangular matrix since it is not a square matrix.

2-7

Now, for review, let us consider the following matrices and identify them properly.

$$\mathbf{M} = \begin{bmatrix} m_1 & m_2 \end{bmatrix}$$

$$\mathbf{N} = \begin{bmatrix} n_{11} & 0 \\ 0 & n_{22} \end{bmatrix}$$

$$\mathbf{P} = \begin{bmatrix} p_{11} \\ p_{21} \\ p_{31} \end{bmatrix}$$

$$\mathbf{Q} = \begin{bmatrix} 1 & 0 & 0 \\ 0 & 2 & 0 \end{bmatrix}$$

$$\mathbf{R} = \begin{bmatrix} 1 & 0 & 0 & 0 \\ 0 & 1 & 0 & 0 \\ 0 & 0 & 1 & 0 \\ 0 & 0 & 0 & 1 \end{bmatrix}$$

$$\mathbf{S} = \begin{bmatrix} 0 & 0 \\ 0 & 0 \end{bmatrix}$$

Identity matrix ____________

Null or zero matrix ________

Diagonal matrix ___________

Row vector ______________

Column vector ___________

The correct identification of the matrices is

Identity matrix: **R**

Null or zero matrix: **S**

Diagonal matrix: **N** and **R**

Row vector: **M**

Column vector: **P**

The unidentified matrix **Q** is of order 2 X 3 and does not qualify as a diagonal matrix since it is not a square matrix.

This is a convenient stopping place in Chapter Two. When you are ready to resume your work simply proceed.

2-8

A special kind of square matrix, which is quite useful in engineering and science applications, is the *symmetric matrix.* As the name implies, it is a square matrix **A** in which the elements of the matrix are such that $a_{ij} = a_{ji}$. That is, the element a_{ij} in the ith row and jth column is equal to the element a_{ji} in the jth row and ith column. A typical symmetric matrix of order 2 might appear as

$$\mathbf{A} = \begin{bmatrix} a_{11} & a_{12} \\ a_{21} & a_{22} \end{bmatrix}$$

where it is required that $a_{12} = a_{21}$.

One example of a symmetric matrix of order 3 is

$$\mathbf{B} = \begin{bmatrix} 4 & 6 & 1 \\ 6 & 8 & 7 \\ 1 & 7 & -2 \end{bmatrix}$$

In this case we note that

$$a_{12} = a_{21} = 6$$

$$a_{13} = a_{31} = 1$$

$$a_{32} = a_{23} = 7$$

Notice that the symmetric matrix appears symmetric about the major diagonal; hence the origin of the name.

Complete the symmetric matrix **C** in the spaces indicated.

$$\mathbf{C} = \begin{bmatrix} 1 & \_\_\_ & 8 & 4 \\ 0 & 3 & \_\_\_ & -2 \\ \_\_\_ & 1 & -6 & 7 \\ 4 & \_\_\_ & 7 & 5 \end{bmatrix}$$

The completed symmetric matrix **C** is

$$\begin{bmatrix} -1 & 0 & 8 & 4 \\ 0 & 3 & 1 & -2 \\ 8 & 1 & -6 & 7 \\ 4 & -2 & 7 & 5 \end{bmatrix}$$

If you had any difficulty in obtaining this answer return to the beginning of the frame and review the definition of a symmetric matrix.

2-9

If k is a scalar number, then a *scalar matrix*, **B**, is obtained when each element of a matrix **A** is multiplied by the scalar number k. Therefore, the scalar matrix **B** is written as

$$\mathbf{B} = \begin{bmatrix} ka_{11} & ka_{12} & \cdots & ka_{1n} \\ ka_{21} & ka_{22} & \cdots & ka_{2n} \\ \cdot & \cdot & \cdot & \cdot \\ \cdot & \cdot & \cdot & \cdot \\ \cdot & \cdot & \cdot & \cdot \\ ka_{m1} & ka_{m2} & \cdots & ka_{mn} \end{bmatrix}$$

This scalar matrix can also be written as

$$\mathbf{B} = k\mathbf{A} = k \begin{bmatrix} a_{11} & \cdots & a_{1n} \\ \cdot & \cdot & \cdot \\ \cdot & \cdot & \cdot \\ \cdot & \cdot & \cdot \\ a_{m1} & \cdots & a_{mn} \end{bmatrix}$$

where $k\mathbf{A}$ implies that the scalar k multiplies each of the elements of the matrix **A**.

Write the scalar matrix **C** which is equal to the identity matrix of order 4 multiplied by the scalar $k = 3$.

$$
\mathbf{C} = 3\mathbf{I} = \begin{bmatrix} 3 & 0 & 0 & 0 \\ 0 & 3 & 0 & 0 \\ 0 & 0 & 3 & 0 \\ 0 & 0 & 0 & 3 \end{bmatrix}
$$

Note that we have obtained a diagonal matrix conveniently written as $3\mathbf{I}$.

2-10

We have defined a matrix as a rectangular array of numbers, and we should note that the numbers may be complex numbers. When a and b are real numbers and $j = \sqrt{-1}$, then $z = a + jb$ is a complex number. The complex numbers $a + jb$ and $a - jb$ are called complex conjugates of each other.

When $\mathbf{A}$ is a matrix with some complex numbers as elements, the matrix obtained from $\mathbf{A}$ by replacing each element by its complex conjugate is called the *conjugate of the matrix* $\mathbf{A}$ and is denoted by $\overline{\mathbf{A}}$.

If

$$
\mathbf{A} = \begin{bmatrix} 3 + j2 & j \\ 4 & 2 - j \end{bmatrix}
$$

then write the conjugate of $\mathbf{A}$.

$$\overline{\mathbf{A}} = \begin{bmatrix} 3-j2 & -j \\ 4 & 2+j \end{bmatrix}$$ is the conjugate of A.

This was obtained by noting that $3 - j2$ is the conjugate of $a_{11} = 3 + j2$. Thus we obtained the conjugate of **A** by obtaining the conjugate of each element of **A**.

If you completed **A** correctly you have completed the chapter and you are ready to review the types of matrices we have considered. Skip to the summary.

If you made an error read the following paragraph, complete the exercise, and then proceed to the summary and review exercises.

Rewritting **A** we have

$$\mathbf{A} = \begin{bmatrix} 3+j2 & +j \\ 4 & 2-j \end{bmatrix} = \begin{bmatrix} a_{11} & a_{12} \\ a_{21} & a_{22} \end{bmatrix}$$

and we wish to obtain $\overline{\mathbf{A}}$. We write the conjugate matrix as

$$\overline{\mathbf{A}} = \begin{bmatrix} \bar{a}_{11} & \bar{a}_{12} \\ \bar{a}_{21} & \bar{a}_{22} \end{bmatrix}$$

Then, since $a_{11} = 3 + j2$, we have $\bar{a}_{11} = 3 - j2$. Similarly, $a_{12} = j$ and therefore $\bar{a}_{12} = -j$. Also, $\bar{a}_{22} = 2 + j$. Finally, $\bar{a}_{21} = a_{21} = 4$ since a_{21} is a real number. Now try the exercise below.

A matrix **B** is given here. Obtain the conjugate of **B** which is written as $\overline{\mathbf{B}}$.

$$\mathbf{B} = \begin{bmatrix} 8 & 6+j2 & 4j \\ 5 & 0 & 1 \\ 1-j & 4 & -2j \end{bmatrix} \qquad \overline{\mathbf{B}} = \begin{bmatrix} 8 & & \\ & 0 & 1 \\ & & \end{bmatrix}$$

$$\overline{\mathbf{B}} = \begin{bmatrix} 8 & 6-j2 & -4j \\ 5 & 0 & 1 \\ 1+j & 4 & 2j \end{bmatrix}$$

SUMMARY

1. A *row vector* is a matrix of order $1 \times n$ written as

$$\mathbf{y} = [y_1 \;\; y_2 \;\; \dots \;\; y_n]$$

2. A *column vector* is a matrix of order $m \times 1$ and is written as

$$\mathbf{u} = \begin{bmatrix} u_1 \\ u_2 \\ \cdot \\ \cdot \\ \cdot \\ u_m \end{bmatrix}$$

3. A *diagonal matrix* contains elements other than zero only on the diagonal of the matrix. Therefore, a square matrix **D** whose elements $d_{ij} = 0$ when $i \neq j$ is called a *diagonal matrix.*

4. A diagonal matrix **I** in which all the elements on the diagonal equal 1 is called the *identity matrix.*

5. A matrix **0**, every element of which is zero, is called a *zero* or *null matrix.*

6. A square matrix **A**, whose elements $a_{ij} = 0$ for $i > j$, is called an *upper triangular matrix.*

7. A square matrix **B** whose elements $b_{ij} = 0$ for $i < j$ is called a *lower triangular matrix.*

8. A square matrix **C** is a *symmetric matrix* when the elements of the matrix are such that $c_{ij} = c_{ji}$.

9. A *scalar matrix* **B** is obtained when each element of a matrix **A** is multiplied by a scalar number k, so that

$$\mathbf{B} = k\mathbf{A} \, .$$

10. The *conjugate of the matrix* **F** is obtained by replacing each element of $\overline{\mathbf{F}}$ by its complex conjugate, and is denoted by $\overline{\mathbf{F}}$.

EXERCISES

Given the following matrices:

$$A = \begin{bmatrix} 4 & 0 \\ 0 & 4 \end{bmatrix} \qquad B = [8 \quad 6 \quad 1] \qquad C = \begin{bmatrix} 1 & 0 & 0 \\ 0 & 1 & 0 \\ 0 & 0 & 1 \end{bmatrix}$$

$$D = \begin{bmatrix} 5 & 3 & 2 \\ 3 & 8 & 4 \\ 2 & 4 & -1 \end{bmatrix} \qquad E = \begin{bmatrix} 8 & 4 & 3 \\ 0 & 7 & 5 \\ 0 & 0 & -1 \end{bmatrix}$$

1. Give the name of each matrix.

   A ______________________ B ______________________

   C ______________________ D ______________________

   E ______________________

2. Write a null matrix of order 2×3.

3. Given:

$$\mathbf{H} \doteq \begin{bmatrix} 7+j & 4j \\ -j & 5 \end{bmatrix}$$

Write the conjugate matrix $\overline{\mathbf{H}}$.

When you have completed this set of exercises, check your answers with those given on page 236.

3 matrix operations

The rules for performing matrix operations such as addition and multiplication have been formulated to make them useful in practical calculations. In this chapter we will use the concepts and definitions we have discussed in the Chapters 1 and 2 to develop several operations in the algebra of matrices.

3-1

First let us consider and explicitly state the useful concept of equality of matrices. Two matrices **A** and **B** are said to be *equal* to each other if, and only if, they have the same order and each element of one is equal to the corresponding element of the other. Therefore, **A** = **B** if, and only if, $a_{ij} = b_{ij}$ for each possible combination of i and j.

If **A** = **B** and **B** = **C** are we able to say that **A** is equal to **C**? Can you show that **A** = **C**? Yes or No? ____________
Give the reason for your answer in your own words before continuing.

Yes, **A** = **C**, because **B** = **C**. You may show that this result satisfied the definition of equal matrices. In order for **A** = **C** it is necessary and sufficient that they are of the same order and that each element of one is equal to the corresponding element of the other. The order of **A** and **C** is equal since the order of **B** and **C** is equal. Also, every element of **A** is equal to the corresponding element of **B**, and similarly, every element of **B** is equal to the corresponding element of **C**. Therefore we can state that every element of **A** equals the corresponding element of **C**. Thus we can say that **A** = **C**.

3-2

Select two equal matrices from the following.

$$A = \begin{bmatrix} 4 & 2 & 1 \\ 6 & 1 & 0 \end{bmatrix} \qquad B = \begin{bmatrix} 4 & 2 \\ 6 & 1 \end{bmatrix}$$

$$C = \begin{bmatrix} 4 & 2 \\ 6 & 1 \end{bmatrix} \qquad D = \begin{bmatrix} 4 & 2 \\ 6 & 1 \\ 0 & 1 \end{bmatrix} \qquad E = \begin{bmatrix} 4 & 6 \\ 6 & 1 \end{bmatrix}$$

Matrix ________________ equals matrix ________________.

You are correct if you stated that matrix **B** equals matrix **C**. If you made an error, recall that for two matrices to be equal the order of the matrices must be equal and each corresponding element must be equal.

3-3

The matrix of order $n \times m$, obtained by interchanging the rows and columns of an $m \times n$ matrix **A**, is called *the transpose* of **A** and is denoted by **A′**. Another common notation for the transpose of **A** is $\mathbf{A}^T$. The a_{ij}th element of the matrix **A** becomes the a_{ji}th element of the transpose matrix **A′**. Note that if the order of the matrix **A** is $m \times$ n then the order of the transposed matrix is $n \times m$. For example, if

$$A = \begin{bmatrix} 3 & 2 \\ 1 & 0 \\ 5 & 1 \end{bmatrix} \qquad \text{of order } 3 \times 2$$

then

$$A' = \begin{bmatrix} 3 & 1 & 5 \\ 2 & 0 & 1 \end{bmatrix} \qquad \text{of order } 2 \times 3$$

Note that the first column of **A** becomes the first row of **A′**. Similarly, the second column of **A** becomes the second row of **A′**

Complete the transpose operation for the matrix **B**.

$$\mathbf{B} = \begin{bmatrix} 5 & 4 & 2 \\ 1 & 0 & 6 \end{bmatrix} \qquad \mathbf{B}' = \begin{bmatrix} 5 & \\ & \\ & \end{bmatrix} \quad \text{of order\_\_\_\_\_\_\_\_}$$

The correct completion of **B′** yields

$$\mathbf{B}' = \begin{bmatrix} 5 & 1 \\ 4 & 0 \\ 2 & 6 \end{bmatrix} \qquad \text{of order } 3 \times 2$$

If you answered correctly, proceed to Frame 3-4. If not, read the following explanation before going on.

Since you made an error let's complete the transpose process step by step. Rewriting **B** we have

$$\mathbf{B} = \begin{bmatrix} 5 & 4 & 2 \\ 1 & 0 & 6 \end{bmatrix}$$

Now b_{11} remains b_{11} in **B′**, and similarly, b_{22} remains b_{22} in **B′**. The first element to interchange is $b_{12} = 4$, which becomes b_{21} in **B′**. In a similar manner $b_{21} = 1$ becomes b_{12} in **B′**. Therefore, so far we have obtained

$$\mathbf{B}' = \begin{bmatrix} 5 & 1 \\ 4 & 0 \\ \_ & \_ \end{bmatrix}$$

with the dashes indicating places to be completed. To complete the matrix we note that $b_{13} = 2$ of **B** becomes $b_{31} = 2$ of **B′**. Finally, b_{32} of **B′** is equal to 6. Note that the order of **B′** is 3 × 2 and that the transpose operation corresponds to the rows of **B** becoming the columns of **B′**.

3-4

Write the transpose of the column vector **x** that has four elements.

$$\mathbf{x}' = \underline{\qquad\qquad\qquad}$$

The transpose of the column vector **x** with four elements is

$$\mathbf{x}' = [x_1 \;\; x_2 \;\; x_3 \;\; x_4]$$

Thus the transpose of a column vector becomes a row vector.

3-5

The transpose of the matrix

$$\mathbf{C} = \begin{bmatrix} 5 & 0 & 2 \\ 2 & 4 & 3 \\ -1 & 8 & 1 \end{bmatrix}$$

is

$$\mathbf{C}' = \begin{bmatrix} 5 & 2 & -1 \\ 0 & 4 & 8 \\ 2 & 3 & 1 \end{bmatrix}$$

The transpose of a lower triangular matrix is an ______________________ ______________________ matrix.

You are correct if you said that the transpose of a lower triangular matrix is an *upper triangular matrix.* (This fact is obtained from the definition of a transpose of a matrix, which requires a_{ij} of the matrix **A** to become a_{ji} of the transposed matrix.)

3-6

The transpose of a symmetric matrix **A**, written **A′**, is equal to__________ That is, using the concept of equal matrices we can write the equation

$$\mathbf{A}' = \_\_\_\_\_\_\_\_\_\_$$

The transpose of a symmetric matrix **A**, written **A′**, is equal to **A**. Thus the equation is written as

$$\mathbf{A}' = \mathbf{A}$$

Recall that the definition of symmetric matrix requires that $a_{ij} = a_{ji}$. Therefore interchanging rows and columns will not change the matrix, and **A′** and **A** are equal.

If you obtained these answers without difficulty, proceed. If you need to review the definition of equality and the concept of a transpose, return to the beginning of this chapter.

ADDITION

3-7

As in arithmetic and ordinary algebra, in matrix algebra we utilize the operation of addition. The addition of two matrices is possible only when they are of the same order. Such matrices are said to be *conformable for addition.* (In this context conformable means suitable or tractable.)

The sum of two matrices **A** and **B**, of order $m \times n$, is another $m \times n$ matrix **C** in which each element is obtained as the sum of the corresponding elements of **A** and **B**. Therefore the sum of **A** and **B** resulting in the matrix **C** is written as

$$\mathbf{C} = \mathbf{A} + \mathbf{B}$$

where element $c_{ij} = a_{ij} + b_{ij}$ for $i = 1, 2, \ldots m$, and $j = 1, 2, \ldots n$. For example,

$$\mathbf{C} = \mathbf{A} + \mathbf{B} = \begin{bmatrix} 4 & 3 & 2 \\ 8 & 1 & -1 \end{bmatrix} + \begin{bmatrix} 0 & 2 & -2 \\ -2 & 7 & 0 \end{bmatrix} = \begin{bmatrix} 4 & 5 & 0 \\ 6 & 8 & -1 \end{bmatrix}$$

The addition of **A** and **A**, written as **C** = **A** + **A**, is equal to a scalar matrix 2**A** so that **C** = 2**A**

True or False? ____________________

True. The addition of **A** and **A**, written as **C** = **A** + **A**, is equal to a scalar matrix 2**A** so that **C** = 2**A**.

We could illustrate this in the following way: Consider a square matrix **A** of order 2. Then, adding **A** + **A**, we have

$$\begin{bmatrix} a_{11} & a_{12} \\ a_{21} & a_{22} \end{bmatrix} + \begin{bmatrix} a_{11} & a_{12} \\ a_{21} & a_{22} \end{bmatrix} = \begin{bmatrix} 2a_{11} & 2a_{12} \\ 2a_{21} & 2a_{22} \end{bmatrix} = 2\begin{bmatrix} a_{11} & a_{12} \\ a_{21} & a_{22} \end{bmatrix}$$

Therefore, we obtain the scalar matrix 2**A**.

If you answered false, review Frame 3-7 before continuing.

3-8

As an exercise, complete the sum of these two matrices:

$$\begin{bmatrix} 8 & 4 & -1 \\ 0 & 1 & 3 \\ 5 & 4 & 8 \end{bmatrix} + \begin{bmatrix} -4 & 6 & 2 \\ 1 & 3 & 7 \\ 5 & 4 & 1 \end{bmatrix} = \begin{bmatrix} 4 & & \\ & & \\ & & 9 \end{bmatrix}$$

The two matrices are of the same order since they are both of order ____________ and therefore are ________________ for addition.

The completed matrix sum is

$$\begin{bmatrix} 4 & 10 & 1 \\ 1 & 4 & 10 \\ 10 & 8 & 9 \end{bmatrix}$$

The two matrices are of the same order since they are both of order *three* and, therefore, are *conformable* for addition.

3-9

In the algebra of numbers, the fact that $a + b = b + a$ for any two numbers is known as the *commutative law of addition.* The fact that $a + (b + c) = (a + b) + c$ for any three scalar numbers, a, b, c, is known as the *associative law of addition.* It is not difficult to see from the manner of performing the addition of matrices that the process is commutative and associative. Therefore, the process of matrix addition can be performed with the matrices added in any order and grouped in any desired manner. In matrix notation we have

Commutative law: $\mathbf{A} + \mathbf{B} = \mathbf{B} + \mathbf{A}$

Associative law: $\mathbf{A} + (\mathbf{B} + \mathbf{C}) = (\mathbf{A} + \mathbf{B}) + \mathbf{C}$

Consider the ijth element of the matrices **A** and **B** of the same order. Using the laws of addition for numbers, show that the commutative law of addition holds for matrices.

In order to prove the commutative law of addition for matrices, we consider the *ij*th element of the matrices **A** and **B**. Then, for the *ij*th element

$$\mathbf{A} + \mathbf{B} \text{ implies } a_{ij} + b_{ij}$$

and

$$\mathbf{B} + \mathbf{A} \text{ implies } b_{ij} + a_{ij}$$

However, we know from the commutative law of numbers that $a_{ij} + b_{ij} = b_{ij} + a_{ij}$. Therefore, since this relation holds for the *ij*th element, it will hold in general, and

$$\mathbf{A} + \mathbf{B} = \mathbf{B} + \mathbf{A}.$$

One may utilize a similar proof for the associative law of addition.

In a similar way, we are able to prove the *cancellation law for addition,* which states that

$$\mathbf{A} + \mathbf{C} = \mathbf{B} + \mathbf{C}$$

if and only if $\mathbf{A} = \mathbf{B}$.

Now let us develop the *subtraction* operation for matrices by considering a scalar matrix. Pause here to consider how *you* would develop this before going on.

SUBTRACTION

3-10

In order to develop the operation of subtraction for matrices, let us reconsider the scalar matrix, which is written as $k\mathbf{A}$. The scalar number may assume any value. In order to facilitate subtraction we will let $k = -1$. Then, we obtain $-\mathbf{A}$, called the *negative* of **A**, which implies that we multiply each element of the matrix **A** by the number -1. Thus, we can write

$$\mathbf{A} + (-\mathbf{A}) = \mathbf{0}$$

where **0** is the null matrix. We can delete the parentheses and write the equivalent equation

$$\mathbf{A} - \mathbf{A} = \mathbf{0}$$

In general, we have the operation of *subtraction* of two matrices as represented by the equation

$$\mathbf{C} = \mathbf{A} - \mathbf{B}$$

which implies that we subtract each element of **B** from the corresponding element of **A**.

For example,

$$\begin{bmatrix} 8 & 4 \\ -2 & 0 \end{bmatrix} - \begin{bmatrix} 5 & 8 \\ 5 & -2 \end{bmatrix} = \begin{bmatrix} 3 & -4 \\ -7 & 2 \end{bmatrix}$$

Show that, if **X**, **A**, and **B** are all of the same order, a solution of the equation $\mathbf{X} + \mathbf{A} = \mathbf{B}$ is $\mathbf{X} = \mathbf{B} - \mathbf{A}$. Write the basic laws of addition and subtraction as you use them.

One way to show that $\mathbf{X} = \mathbf{B} - \mathbf{A}$ is to substitute this solution into the left-hand side of the relation

$$\mathbf{X} + \mathbf{A} = \mathbf{B}$$

and show that the result, after suitable algebraic manipulation, is indeed equal to **B**. Substituting $\mathbf{X} = \mathbf{B} - \mathbf{A}$, we have

$$\begin{aligned} (\mathbf{B} - \mathbf{A}) + \mathbf{A} &= \mathbf{B} + (-\mathbf{A} + \mathbf{A}) && \text{associative law} \\ &= \mathbf{B} + \mathbf{0} && \text{subtraction rule} \\ &= \mathbf{B} \end{aligned}$$

which indeed is equal to the right side of the original relationship, $\mathbf{X} + \mathbf{A} = \mathbf{B}$.

3-11

Consider the relationship

$$(\mathbf{A} - \mathbf{B}) + \mathbf{C} = \mathbf{A} - (\mathbf{B} + \mathbf{C})$$

Is this a correct relationship? Yes or No?_______ State the reason for your answer.

No, it is not a correct relationship since the left side of the equality sign is not equal to the right side. Therefore

$$(\mathbf{A} - \mathbf{B}) + \mathbf{C} \neq \mathbf{A} - (\mathbf{B} + \mathbf{C})$$

This can be readily shown by utilizing the operation of subtraction and the associative law of addition. Clearly, the right side of the equation can be written as

$$(\mathbf{A} - \mathbf{B}) - \mathbf{C}$$

which is not equal, in general, to the left side, **A** – (**B** + **C**).

3-12

Complete the operation (**F** – **H**) + **M** when

$$\mathbf{F} = \begin{bmatrix} 6 & 2 \\ 0 & -1 \end{bmatrix}; \quad \mathbf{H} = \begin{bmatrix} 3 & 1 \\ -2 & 2 \end{bmatrix}; \quad \mathbf{M} = \begin{bmatrix} 1 & 6 \\ -3 & 4 \end{bmatrix}$$

$$\begin{matrix} \mathbf{F} & - & \mathbf{H} & & \\ \begin{bmatrix} 6 & 2 \\ 0 & -1 \end{bmatrix} & - & \begin{bmatrix} 3 & 1 \\ -2 & 2 \end{bmatrix} & = & \begin{bmatrix} 3 & 1 \\ +2 & -3 \end{bmatrix} \end{matrix}$$

and then

$$\begin{matrix} (\mathbf{F}-\mathbf{H}) & + & \mathbf{M} & & \\ \begin{bmatrix} 3 & 1 \\ 2 & -3 \end{bmatrix} & + & \begin{bmatrix} 1 & 6 \\ -3 & 4 \end{bmatrix} & = & \begin{bmatrix} 4 & 7 \\ -1 & 1 \end{bmatrix} \end{matrix}$$

This is a convenient stopping place in Chapter Three.

MATRIX MULTIPLICATION

3-13

When matrices are used to solve a set of linear algebraic equations, a multiplication operation is particularly useful. Therefore, *matrix multiplication* is defined so it is particularly suitable for linear algebraic equations in matrix form.

Initially, let us consider the multiplication of a row vector **x** by a column vector **y**. It is required that they be *conformable for multiplication,* which implies that they both contain n elements. Then the matrix multiplication is written as

$$\mathbf{xy} = [x_1 \quad x_2 \quad \ldots \quad x_n] \begin{bmatrix} y_1 \\ y_2 \\ \cdot \\ \cdot \\ \cdot \\ y_n \end{bmatrix}$$

The matrix multiplication of a $1 \times n$ row vector and an $n \times 1$ column vector is defined by a resulting 1×1 matrix **C** such that

$$\mathbf{C} = [x_1y_1 + x_2y_2 + \ldots + x_ny_n]$$

That is, the result of a row-column multiplication is a scalar number. For example, the multiplication of a specific **x** and **y** is

$$\mathbf{xy} = [2 \quad 5 \quad -2] \begin{bmatrix} 3 \\ -3 \\ 4 \end{bmatrix} = [2(3) + 5(-3) + (-2)(4)] = -17$$

Find the product **w′z** where

$$\mathbf{w} = \begin{bmatrix} 8 \\ 6 \\ 1 \end{bmatrix} \qquad \mathbf{z} = \begin{bmatrix} 2 \\ -1 \\ 3 \end{bmatrix}$$

You are correct if you obtained the product **w′z** as

$$\mathbf{w'z} = [8 \quad 6 \quad 1] \begin{bmatrix} 2 \\ -1 \\ 3 \end{bmatrix} = [8(2) + 6(-1) + 1(3)] = 13$$

Recall, from Frame 3-5 that the transpose of a column vector results in a row vector.

Note that the operation is row by column, so that each element of the row is multiplied into the corresponding element of the column and then the products are added. This multiplication approach is extended to $n \times m$ matrices to obtain a definition of matrix multiplication.

Let us consider the multiplication of a matrix **A** of order 3×2 by a matrix **B** of order 2×2. This product is written as

(3-2)

$$\mathbf{C} = \mathbf{AB} = \begin{bmatrix} a_{11} & a_{12} \\ a_{21} & a_{22} \\ a_{31} & a_{32} \end{bmatrix} \begin{bmatrix} b_{11} & b_{12} \\ b_{21} & b_{22} \end{bmatrix}$$

In order for the multiplication to be accomplished, it is necessary for the number of columns of **A** (two) to be equal to the number of rows of **B**, which is also two. Then *matrix multiplication* consists of multiplying a specific row of **A** by a column of **B**, resulting in one element of **C**, where in this case

$$c_{ij} = a_{i1}b_{1j} + a_{i2}b_{2j}$$

That is, each column of **A** is multiplied once into each column of **B**, resulting in one element of **C**. Therefore, for illustrative purposes, let us consider the product of equation (3-1).

$$\mathbf{C} = \mathbf{AB} = \begin{bmatrix} a_{11} & a_{12} \\ a_{21} & a_{22} \\ a_{31} & a_{32} \end{bmatrix} \begin{bmatrix} b_{11} & b_{12} \\ b_{21} & b_{22} \end{bmatrix}$$

The first element of the **C** matrix is

$$c_{11} = a_{11}b_{11} + a_{12}b_{21}$$

which is obtained by multiplying the first row of **A** by the first column of **B**. Remember, $c_{ij} = a_{i1}b_{1j} + a_{i2}b_{2j}$ for this specific illustration.

The second element of **C**, c_{12}, is then

$$c_{12} = a_{11}\ \_\_\_ + \_\_\_\ b_{22}$$

The second element of **C**, c_{12}, is

$$c_{12} = a_{11}b_{12} + a_{12}b_{22}$$

which is obtained by multiplying the first row of **A** by the second column of **B**.

3-15

The element c_{32} is

If you said that $c_{32} = a_{31}b_{12} + a_{32}b_{22}$, you are doing very well. This results from multiplying the third row of **A** by the second column of **B**, which is implied by the subscript of the element $c_{ij} = c_{32}$.

3-16

Now let us state a general definition of *matrix multiplication.* Let **A** be an $m \times p$ matrix and **B** be a $p \times n$ matrix. The *product* **AB** is then defined to be the $m \times n$ matrix whose element in the ith row and jth column is found by multiplying corresponding elements of the ith row of **A** and of the jth column of **B** and then adding the resulting product terms. Therefore,

$$\mathbf{C} = \mathbf{AB}$$

where

$$c_{ij} = a_{i1}b_{1j} + a_{i2}b_{2j} + \dots + a_{ip}b_{pj} \tag{3-2}$$

and i and j assume values from 1 to m and n respectively. The resulting product, **C**, has the same number of rows as **A** and the same number of columns as **B**. Furthermore, we note that in order to carry out the multiplication, the number of columns of **A** must be equal to the number of rows of **B**. Then the matrix **A** is said to be *conformable to* **B** and the multiplication **AB** exists.

The element c_{23} in the product of two square matrices **A** and **B** of order 3 is c_{23} =

$c_{23} = a_{21}b_{13} + a_{22}b_{23} + a_{23}b_{33}$ is the correct answer.

3-17

Using a summation notation, equation (3-2) can be written as

$$c_{ij} = \sum_{k=1}^{p} a_{ik}b_{kj}$$

where we recall that p was the number of columns of **A** and the number of rows of **B**. The summation sign Σ indicates that we should let $k = 1$ to p and sum the resulting terms of the expansion.

Write the expansion for the element c_{22} in the product of squares matrices of order 2 using the summation notation as well as the completed expansion.

$$c_{12} = \sum_{k=} \_\_\_\_\_\_ = \_\_\_\_\_\_\_\_\_\_$$

$c_{12} = \sum_{k=1}^{2} a_{1k}b_{k2} = a_{11}b_{12} + a_{12}b_{22}$ is the correct answer

3-18

Now let us consider an example of matrix multiplication. Let us determine the product **AB** when

$$\mathbf{A} = \begin{bmatrix} 4 & -1 & 0 \\ 2 & 3 & 8 \end{bmatrix} \; ; \; \mathbf{B} = \begin{bmatrix} 3 & 4 \\ 2 & 1 \\ 0 & 2 \end{bmatrix}$$

We must examine the order of each matrix to determine if they are conformable for multiplication. Matrix **A** is of order 2 X 3; matrix **B** is of order 3 X 2. Since the number of columns of **A** is equal to the number of rows of **B**, the matrices are conformable. The matrix **C** resulting from the product will be of order 2 X 2.

The first element of **C** is c_{11} and is obtained by multiplying the first row of **A** by the first column of **B** so that

$$c_{11} = 4(3) + (-1)(2) + (0)(0) = 10$$

Carrying out the remaining row-column operations we have

$$\mathbf{C} = \mathbf{AB} = \begin{bmatrix} (4(3) - 1(2) + 0(0)) & (4(4) - 1(1) + 0(2)) \\ (2(3) + 3(2) + 8(0)) & (2(4) + 3(1) + 8(2)) \end{bmatrix}$$

$$= \begin{bmatrix} 10 & 15 \\ 12 & 27 \end{bmatrix}$$

Can the product **BA** be obtained for the matrices **A** and **B** given in this example? Yes or No?______________ State the reason for your answer.

Yes, the product **BA** can be obtained for these matrices. Now, the question of whether **BA** can be obtained turns on the point of conformability of **B** and **A**. The matrix **B** is of order 3 X 2 and the matrix **A** is of order 2 X 3. Recall that the multiplication operation calls for the multiplication of a row of the first matrix, **B** in this case, by a column of the second matrix **A**. Therefore, we require the number of columns of **B** to be equal to the number of rows of **A** in order for the matrices to be conformable. The matrices **B** and **A** satisfy this requirement since **B** has two columns and **A** has two rows. Therefore, the multiplication can be accomplished.

3-19

Is the product **xy** of a column and row vector conformable for multiplication where

$$\mathbf{C} = \mathbf{xy} = \begin{bmatrix} x_1 \\ x_2 \end{bmatrix} \quad [y_1 \quad y_2]$$

Yes or No? ______________

Why?

If conformable, complete the operation

$$\mathbf{C} = \mathbf{xy} =$$

Yes

$$C = xy = \begin{bmatrix} x_1 \\ x_2 \end{bmatrix} [y_1 \;\; y_2]$$

is conformable for multiplication because **x** has one column and **y** has one row. The resulting matrix product is

$$C = xy = \begin{bmatrix} x_1 \\ x_2 \end{bmatrix} [y_1 \;\; y_2] = \begin{bmatrix} (x_1y_1) & (x_1y_2) \\ (x_2y_1) & (x_1y_2) \end{bmatrix}$$

where **C** is a matrix of order 2 X 2. The size of **C** can be verified by considering the order of **x** which is 2 X 1, and that of **y**, which is 1 X 2.

3-20

Calculate the products **AB** and **BA** for the following matrices:

$$A = \begin{bmatrix} 1 & 2 \\ 3 & 4 \end{bmatrix} \; ; \quad B = \begin{bmatrix} 1 & 1 \\ 2 & 0 \end{bmatrix}$$

(a) Are the matrices conformable for multiplication as the products **AB** and **BA**? (See Frame 3-18 for review, if necessary.) ______________________

(b) Matrix multiplication is not commutative in most cases. That is, usually **AB** ≠ **BA**. Is the matrix multiplication commutative in this case?

____

(a) The matrices **A** and **B** are both square matrices of order 2. Therefore they both have the same number of rows and columns and are conformable for multiplication as the products **AB** and **BA**. The products are

$$\mathbf{AB} = \begin{bmatrix} 5 & 1 \\ 11 & 3 \end{bmatrix} \quad \text{and} \quad \mathbf{BA} = \begin{bmatrix} 4 & 6 \\ 2 & 4 \end{bmatrix}$$

(b) Clearly, the matrices are not commutative for multiplication since $\mathbf{AB} \neq \mathbf{BA}$.

3-21

The matrix product **AI** is commutative where **I** is the identity matrix, and **A** and **I** are square matrices of the same order.

Can you show that $\mathbf{AI} = \mathbf{IA}$?

Yes, it is easy to show that **AI** = **IA**. For example, for a square matrix of order 2 we have

$$\mathbf{AI} = \begin{bmatrix} a_{11} & a_{12} \\ a_{21} & a_{22} \end{bmatrix} \begin{bmatrix} 1 & 0 \\ 0 & 1 \end{bmatrix} = \begin{bmatrix} a_{11} & a_{12} \\ a_{21} & a_{22} \end{bmatrix} = \mathbf{A}$$

Of course, in order to multiply **A** times **I**, the identity matrix must also be of order 2. Then, we obtain

$$\mathbf{IA} = \begin{bmatrix} 1 & 0 \\ 0 & 1 \end{bmatrix} \begin{bmatrix} a_{11} & a_{12} \\ a_{21} & a_{22} \end{bmatrix} = \begin{bmatrix} a_{11} & a_{12} \\ a_{21} & a_{22} \end{bmatrix} = \mathbf{A}$$

Irrespective of the order of the square matrix **A** we have a commutative product and **AI** = **IA** = **A**.

3-22

Calculate the product **AB** when

$$\mathbf{A} = \begin{bmatrix} 1 & 2 \\ 3 & 6 \end{bmatrix} \quad \text{and} \quad \mathbf{B} = \begin{bmatrix} 2 & -4 \\ -1 & 2 \end{bmatrix}$$

The product is

$$\mathbf{AB} = \begin{bmatrix} 0 & 0 \\ 0 & 0 \end{bmatrix} = \mathbf{0}$$

where **0** is the null matrix. Thus, we find that the product **AB** = **0** does not imply that **A** or **B** are null matrices or that any element of **A** or **B** is zero.

3-23

It is also true that the relation **AB** = **AC** does not necessarily imply that **B** = **C**. Remember, the matrix multiplication operation is not similar to the multiplication of numbers and we cannot directly utilize our prior experience with the multiplication of numbers.

In order to show **AB** = **AC** does not imply **C** = **B** we shall use **A** and **B** from Frame 3-22, which gave **AB** = **0**. Find **AC** if

$$\mathbf{C} = \begin{bmatrix} -6 & 10 \\ 3 & -5 \end{bmatrix}$$

Note again that **AC** = **0** while **C** ≠ **B.**

SUMMARY

In this chapter we have considered several important matrix operations. We have studied the matrix operations of addition, subtraction, and multiplication. Furthermore we have defined the equality of matrices and the transpose operation for a matrix.

In summary, the important definitions and properties of these matrix operations are

1. Two matrices **A** and **B** are said to be equal if and only if they have the same order, and each element of one is equal to the corresponding element of the other.

2. The matrix of order $n \times m$ obtained by interchanging the rows and columns of an $m \times n$ matrix **A** is called the transpose of **A** and is denoted by **A′**.

3. The sum of two matrices **A** and **B** of order $m \times n$ is another matrix **C** in which each element is obtained as the sum of the corresponding elements of **A** and **B**.

4. The laws of addition are

   (a) Commutative law: **A + B = B + A**

   (b) Associative law: **A + (B + C) = (A + B) + C**

   (c) Cancellation law: **A + C = B + C** if and only if **A = B**

5. The subtraction of two matrices of the same order, represented by the equation **C = A − B**, implies that one subtracts each element of **B** from the corresponding element of **A**.

6. When **A** is an $m \times p$ matrix and **B** is a $p \times n$ matrix, then **A** and **B** are said to be conformable for multiplication as **AB**. The product **AB** is an $m \times n$ matrix in which the element in the ith row and jth column is found by multiplying corresponding elements of the ith row of **A** and the jth column of **B** and then adding the resulting terms. Therefore, **C = AB** where

$$c_{ij} = \sum_{k=1}^{p} a_{ik} b_{kj}$$

EXERCISES

1. Given the following relationships for the matrices **A, B, C, F**, and **G**:

$$\mathbf{A} + \mathbf{B} = \mathbf{F}$$

and

$$\mathbf{B} + \mathbf{C} = \mathbf{G}$$

What law is illustrated by the following equation?

$$\mathbf{A} + \mathbf{G} = \mathbf{F} + \mathbf{C}$$

2. Determine the product $\mathbf{xy}'$ where

$$\mathbf{x} = \begin{bmatrix} 2 \\ 1 \end{bmatrix} \quad \text{and} \quad \mathbf{y} = \begin{bmatrix} 5 \\ 2 \end{bmatrix}$$

3. When

$$\mathbf{A} = \begin{bmatrix} 8 & 3 \\ 2 & 1 \end{bmatrix} \quad \text{and} \quad \mathbf{B} = \begin{bmatrix} 4 & -2 \\ 0 & 1 \end{bmatrix}$$

determine

(a) $\mathbf{A} + \mathbf{B}$

(b) $\mathbf{A} - \mathbf{B}$

4. Obtain the product **AB** when

$$\mathbf{A} = \begin{bmatrix} 8 & 6 & 1 \\ \\ 0 & 2 & 3 \end{bmatrix} \quad \text{and} \quad \mathbf{B} = \begin{bmatrix} 0 & 1 \\ 2 & 1 \\ 1 & 0 \end{bmatrix}$$

5. Does the product **AB** equal the product **BA** in general?

Check your answers against those given on page 237.

4 linear equations and determinants

4-1

Matrix algebra is particularly useful for the solution of sets of simultaneous linear algebraic equations that arise in engineering and science. A set of two simultaneous linear equations with two unknowns, x_1 and x_2, could be written as

(4-1)
$$a_{11}x_1 + a_{12}x_2 = b_1$$
$$a_{21}x_1 + a_{22}x_2 = b_2$$

where the a_{ij} are the known coefficients of the equations and the b_i are known constants. A specific numerical example might yield

(4-2)
$$3x_1 + 2x_2 = 8$$
$$4x_1 + 6x_2 = 10$$

where x_1 and x_2 are unknown.

For review, solve the unknown x_1 in the linear equation

$$3x_1 + 2 = 11$$

$x_1 = 3$ is the answer since

$$3x_1 = 11 - 2$$
$$= 9 \qquad \text{or } x_1 = \frac{9}{3}$$

4-2

We now want to develop a procedure for solving a set of simultaneous linear equations. Rewriting equation (4-1) we have

(4-3)
$$a_{11}x_1 + a_{12}x_2 = b_1$$
$$a_{21}x_1 + a_{22}x_2 = b_2$$

This set of equations can be represented by a matrix equation when we recall the multiplication process, so that

$$\begin{bmatrix} a_{11} & a_{12} \\ a_{21} & a_{22} \end{bmatrix} \begin{bmatrix} x_1 \\ x_2 \end{bmatrix} = \begin{bmatrix} (a_{11}x_1 + a_{12}x_2) \\ (a_{21}x_1 + a_{22}x_2) \end{bmatrix}$$

represents the left side of the equation.

As usual we let

$$\mathbf{A} = \begin{bmatrix} a_{11} & a_{12} \\ a_{21} & a_{22} \end{bmatrix} \quad \text{and} \quad \mathbf{x} = \begin{bmatrix} x_1 \\ x_2 \end{bmatrix}$$

x is a ______________________ vector of order ______________ and the product **Ax** results in a matrix of order ______________________

x is a column vector of order 2 X 1 and the product **Ax** results in a matrix of order 2 X 1.

4-3

Recall from Chapter 3 that the conformable multiplication of a matrix by a column vector yields a column vector.

The right side of equation (4-3) is represented by a column vector

$$\mathbf{b} = \begin{bmatrix} b_1 \\ b_2 \end{bmatrix}$$

Therefore the proper equality holds and we have the matrix equation

$$\mathbf{Ax} = \mathbf{b}$$

Rewriting equation (4-2) we have

$$3x_1 + 2x_2 = 8$$

$$4x_1 + 6x_2 = 10$$

Complete the matrix equation $\mathbf{Ax} = \mathbf{b}$

$$\begin{bmatrix} \quad & 2 \\ \quad & \quad \end{bmatrix} \begin{bmatrix} x_1 \\ \quad \end{bmatrix} = \begin{bmatrix} \quad \\ 10 \end{bmatrix}$$

$$\begin{bmatrix} 3 & 2 \\ 4 & 6 \end{bmatrix} \begin{bmatrix} x_1 \\ x_2 \end{bmatrix} = \begin{bmatrix} 8 \\ 10 \end{bmatrix}$$

If you answered correctly, skip to Frame 4-5. If your answer was incorrect be sure to read Frame 4-4.

4-4

Since you made an error, let's complete the multiplication $\mathbf{Ax}$ and show that it leads to the original set of equations.

$$\mathbf{Ax} = \begin{bmatrix} 3 & 2 \\ 4 & 6 \end{bmatrix} \begin{bmatrix} x_1 \\ x_2 \end{bmatrix} = \begin{bmatrix} (3x_1 + 2x_2) \\ (4x_1 + 6x_2) \end{bmatrix}$$

and $\mathbf{Ax} = \mathbf{b}$ yields

$$\begin{bmatrix} (3x_1 + 2x_2) \\ (4x_1 + 6x_2) \end{bmatrix} = \begin{bmatrix} 8 \\ 10 \end{bmatrix}$$

or equating elements we have

$$3x_1 + 2x_2 = 8 \qquad 4x_1 + 6x_2 = 10$$

Now write the following set of equations in matrix form **Ax = b**

$$2x_1 + 6x_2 = 5 \qquad 3x_1 + 2x_2 = 6$$

$$\begin{bmatrix} 2 & 6 \\ 3 & 2 \end{bmatrix} \begin{bmatrix} x_1 \\ x_2 \end{bmatrix} = \begin{bmatrix} 5 \\ 6 \end{bmatrix}$$

is the correct answer. It may be useful for you to check this answer by completing the multiplication.

4-5

A system of linear equations in three unknowns would be written as

$$a_{11}x_1 + a_{12}x_2 + a_{13}x_3 = b_1$$

$$a_{21}x_1 + a_{22}x_2 + a_{23}x_3 = b_2$$

$$a_{31}x_1 + a_{32}x_2 + a_{33}x_3 = b_3$$

and in matrix form we have

$$\begin{bmatrix} \quad & \quad & \quad \\ \quad & \quad & \quad \\ \quad & \quad & \quad \end{bmatrix} \begin{bmatrix} \quad \\ \quad \\ \quad \end{bmatrix} = \begin{bmatrix} \quad \\ \quad \\ \quad \end{bmatrix}$$

or

$$\mathbf{Ax} = \mathbf{b}$$

Be sure to fill in the matrices before continuing.

$$\begin{bmatrix} a_{11} & a_{12} & a_{13} \\ a_{21} & a_{22} & a_{23} \\ a_{31} & a_{32} & a_{33} \end{bmatrix} \begin{bmatrix} x_1 \\ x_2 \\ x_3 \end{bmatrix} = \begin{bmatrix} b_1 \\ b_2 \\ b_3 \end{bmatrix}$$

or

$$\mathbf{Ax} = \mathbf{b}$$

4-6

Clearly, we can use the matrix equation $\mathbf{Ax} = \mathbf{b}$ to represent a set of n equations in n unknowns. The matrix notation for n equation with n unknowns is

$$\begin{bmatrix} a_{11} & a_{12} & \cdots & a_{1n} \\ a_{21} & a_{22} & \cdots & a_{2n} \\ \cdot & \cdot & & \cdot \\ \cdot & \cdot & & \cdot \\ \cdot & \cdot & & \cdot \\ a_{n1} & a_{n2} & \cdots & a_{nn} \end{bmatrix} \begin{bmatrix} x_1 \\ x_2 \\ \cdot \\ \cdot \\ \cdot \\ x_n \end{bmatrix} = \begin{bmatrix} b_1 \\ b_2 \\ \cdot \\ \cdot \\ \cdot \\ b_n \end{bmatrix}$$

The dots indicate the presence of consistent symbols.

Write the following set of equations in matrix form using the matrices below

$$3x_1 + 2x_3 + 4x_4 = 6$$

$$2x_2 + 6x_4 = 8$$

$$5x_1 + 2x_2 + x_4 = 10$$

$$\mathbf{Ax} = \mathbf{b}$$

$$\begin{bmatrix} 3 & & & \\ & & & \\ & & & 1 \end{bmatrix} \begin{bmatrix} x_1 \\ \\ \\ \\ \end{bmatrix} = \begin{bmatrix} \\ \\ \\ \end{bmatrix}$$

Noting that we have three equations in four unknowns we obtain an **A** matrix of order 3 X 4 and

$$\begin{bmatrix} 3 & 0 & 2 & 4 \\ 0 & 2 & 0 & 6 \\ 5 & 2 & 0 & 1 \end{bmatrix} \begin{bmatrix} x_1 \\ x_2 \\ x_3 \\ x_4 \end{bmatrix} = \begin{bmatrix} 6 \\ 8 \\ 10 \end{bmatrix}$$

Completing the multiplication process will verify that one obtains the original set of equations.

4-7

The matrix equation $\mathbf{Ax} = \mathbf{b}$ is called a *nonhomogeneous* equation. The matrix equation $\mathbf{Ax} = \mathbf{0}$ is called a *homogeneous* equation where $\mathbf{0}$ is the zero column matrix.

We are interested in obtaining the solution of the nonhomogeneous equation by the use of determinants.

$$\begin{bmatrix} 3 & 2 \\ 1 & 4 \end{bmatrix} \begin{bmatrix} x_1 \\ x_2 \end{bmatrix} = \begin{bmatrix} 8 \\ 0 \end{bmatrix}$$

is a homogeneous equation. True or False? ____________________

False.

$$\mathbf{Ax} = \begin{bmatrix} 8 \\ 0 \end{bmatrix}$$

is nonhomogeneous since the **b** matrix is not a zero matrix. Recall that all the elements are equal to zero in a zero or null matrix.

DETERMINANTS

4-8

The determinant of an $n \times n$ matrix **A** is a unique scalar number that we associate with **A** by a well-defined calculation rule. Determinants serve a useful role in the solution of simultaneous equations as well as in other matrix operations. First let us consider the *determinant* of a 2×2 matrix **A** where

$$\mathbf{A} = \begin{bmatrix} a_{11} & a_{12} \\ a_{21} & a_{22} \end{bmatrix}$$

We represent the *determinant of a square matrix* **A** as

$$\det \mathbf{A} = \begin{vmatrix} a_{11} & a_{12} \\ a_{21} & a_{22} \end{vmatrix}$$

where the straight lines without the usual corners indicate that it is a determinant. For the 2 X 2 matrix we have

$$\det A = \begin{vmatrix} a_{11} & a_{12} \\ a_{21} & a_{22} \end{vmatrix} = a_{11}a_{22} - a_{12}a_{21}$$

In schematic form the process can be illustrated with arrows as

$$\det A = \begin{vmatrix} \overset{(+)}{a_{11}} & \overset{(-)}{a_{12}} \\ a_{21} & a_{22} \end{vmatrix}$$

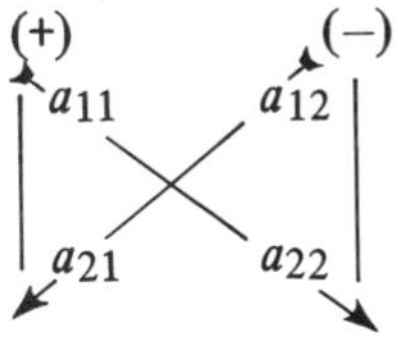

Evaluate the determinant of the matrix

$$A = \begin{bmatrix} 3 & 2 \\ 6 & 5 \end{bmatrix}$$

27 See below.

det A =

3 See the answer above Frame 4-10 on page 70.

You said that det $\begin{bmatrix} 3 & 2 \\ 6 & 5 \end{bmatrix}$ equals 27, which is incorrect.

Undoubtedly you added the two products $a_{11}a_{22}$ and $a_{12}a_{21}$. Remember

$$\det A = a_{11}a_{22} - a_{12}a_{21}$$

and therefore

$$\det \begin{bmatrix} 3 & 2 \\ 6 & 5 \end{bmatrix} = 3(5) - 2(6) = 3$$

4-9

Find the determinant of $\mathbf{A} = \begin{bmatrix} 2 & 1 \\ 2 & 0 \end{bmatrix}$

You are correct if you stated that

$$\det \begin{bmatrix} 2 & 1 \\ 2 & 0 \end{bmatrix} = 2(0) - 2(1) = -2$$

Now continue with Frame 4-10.

From Frame 4-8.

Correct, det $\mathbf{A} = 3$.

4-10

The determinant of a 3 X 3 square matrix **A** is

$$\det \begin{bmatrix} a_{11} & a_{12} & a_{13} \\ a_{21} & a_{22} & a_{23} \\ a_{31} & a_{32} & a_{33} \end{bmatrix} = a_{11}a_{22}a_{33} + a_{12}a_{23}a_{31} + a_{13}a_{21}a_{32} - a_{13}a_{22}a_{31} - a_{11}a_{23}a_{32} - a_{12}a_{21}a_{33}$$

This expression can be represented schematically by arrows

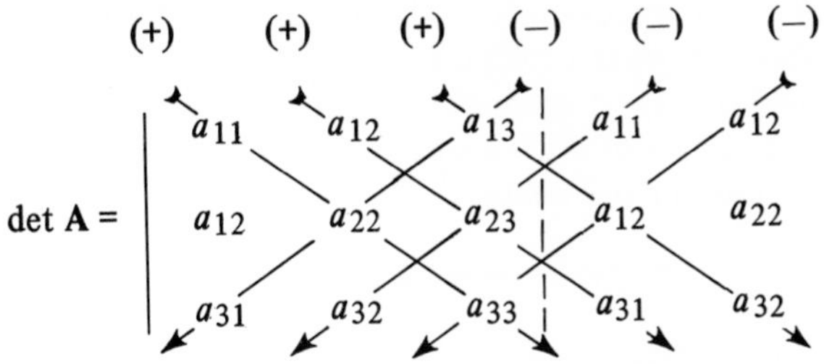

where the two columns are repeated to enable the procedure to be accomplished.

Evaluate

$$\det \begin{bmatrix} 6 & 2 & 1 \\ 3 & 0 & 1 \\ 8 & 2 & 3 \end{bmatrix} =$$

$$\det \begin{bmatrix} 6 & 2 & 1 \\ 3 & 0 & 1 \\ 8 & 2 & 3 \end{bmatrix} = 6(0)3 + 2(1)8 + 1(3)2 - 1(0)8 - 6(1)2 - 2(3)3 = -8$$

This process for evaluating the determinant of **A** is suitable for square matrices of order 2 or 3. For higher order matrices it is necessary to use other procedures.

4-11

A *minor* of a determinant is another determinant formed by removing an equal number of rows and columns from the original determinant. The *order* of the minor indicates the number of rows (or columns) in the minor.

For a determinant of order 3 we have nine second-order minors obtained by deleting one row and one column.

$$\det \mathbf{A} = \begin{vmatrix} a_{11} & a_{12} & a_{13} \\ a_{21} & a_{22} & a_{23} \\ a_{31} & a_{32} & a_{33} \end{vmatrix}$$

The three minors obtained by deleting the first row are

$$|M_{11}| = \begin{vmatrix} a_{22} & a_{23} \\ a_{32} & a_{33} \end{vmatrix} \qquad |M_{12}| = \begin{vmatrix} a_{21} & a_{23} \\ a_{31} & a_{33} \end{vmatrix}$$

$$|M_{13}| = \begin{vmatrix} a_{21} & a_{22} \\ a_{31} & a_{32} \end{vmatrix}$$

where $|M_{ij}|$ refers to the minor of $|\mathbf{A}|$ obtained by removing the ith row and jth column.

Let us evaluate the three minors obtained from the first row of the determinant

$$\det \mathbf{A} = \begin{vmatrix} 6 & 2 & 1 \\ 3 & 0 & 1 \\ 8 & 2 & 3 \end{vmatrix}$$

The first minor $|M_{11}|$ which is a determinant itself, is obtained by removing the first row and first column from $|\mathbf{A}|$ so that

$$|M_{11}| = \begin{vmatrix} 0 & 1 \\ 2 & 3 \end{vmatrix} = 0(3) - 1(2) = -2$$

Similarly,

$$|M_{12}| = \begin{vmatrix} 3 & 1 \\ 8 & 3 \end{vmatrix} = 3(3) - 1(8) = 1$$

which is obtained by removing the first row and second column of $|\mathbf{A}|$.

Evaluate the minor $|M_{13}|$

$$|M_{13}| = \begin{vmatrix} 3 & 0 \\ 8 & 2 \end{vmatrix} = 3(2) - 0(8) = 6$$

4-12

The minor $|M_{22}|$ is obtained by removing the second row and the second column so that

$$|M_{22}| = \begin{vmatrix} 6 & 2 & 1 \\ 3 & 0 & 1 \\ 8 & 2 & 3 \end{vmatrix} = \begin{vmatrix} 6 & 1 \\ 8 & 3 \end{vmatrix} = 6(3) - 1(8) = 10$$

The *cofactor* of an element of a determinant is found by giving an appropriate sign (plus or minus) to the minor obtained when the row and column corresponding to that element are removed. In other words the signed minor

$$(-1)^{i+j}\ |M_{ij}|$$

is called the cofactor of a_{ij} and is denoted by α_{ij} so that

$$\alpha_{ij} = (-1)^{i+j}\ |M_{ij}|$$

Therefore considering the foregoing example,

$$\alpha_{22} = (-1)^4 \, |M_{22}| = 10$$

What is α_{32} for

$$\det \mathbf{A} = \begin{vmatrix} 6 & 2 & 1 \\ 3 & 0 & 1 \\ 8 & 2 & 3 \end{vmatrix} ?$$

$$\alpha_{32} = (-1)^{3+2} \, |M_{32}|$$

$$= (-1)^5 \begin{vmatrix} 6 & 1 \\ 3 & 1 \end{vmatrix} = (-1)\,(6(1) - 1(3)) = -3$$

If you would like further practice, turn to Frame 4-14; otherwise, continue here.

4-13

It is possible to evaluate any determinant through the use of cofactors of a selected row or column. This method is called the *expansion by cofactors* and consists of

1. selecting a row (or column) of the determinant
2. multiplying each element in the row (or column) by its cofactor
3. obtaining the sum of the products obtained in step 2 which equals the value of the determinant. When we select a row this procedure is represented by the equation

(4-4) $$\det \mathbf{A} = a_{i1}\alpha_{i1} + a_{i2}\alpha_{i2} + \ldots + a_{in}a_{in}$$

$$= \sum_{j=1}^{n} a_{ij}\alpha_{ij}$$

where i is the row to be selected.

Therefore, for a 3 order determinant, selecting the first row we have

$$\det \mathbf{A} = \begin{vmatrix} a_{11} & a_{12} & a_{13} \\ a_{21} & a_{22} & a_{23} \\ a_{31} & a_{32} & a_{33} \end{vmatrix} = a_{11}\alpha_{11} + a_{12}\alpha_{12} + a_{13}\alpha_{13}$$

$$= a_{11} \; |M_{11}| + a_{12}(-1) \; |M_{12}| + a_{13} \; |M_{13}|$$

Skip to the bottom of 76.

4-14

Consider

$$\det \begin{vmatrix} 1 & 3 & 0 \\ 8 & 2 & 1 \\ 0 & 4 & 2 \end{vmatrix}$$

Evaluate the cofactor α_{23}. Then we have

$$\alpha_{23} = (-1)^{2+3} \begin{vmatrix} 1 & 3 \\ 0 & 4 \end{vmatrix} = (-1)\,(1(4) - 3(0)) = -4$$

Evaluate the cofactor α_{12}

$$\alpha_{12} = (-1)^3 \begin{vmatrix} 8 & 1 \\ 0 & 2 \end{vmatrix} = (-1)(16) = -16$$

Continue now with Frame 4-13.

From page 75.

Let us use this expansion procedure to evaluate the determinant previously considered.

$$\det \begin{vmatrix} 6 & 2 & 1 \\ 3 & 0 & 1 \\ 8 & 2 & 3 \end{vmatrix} = 6 \begin{vmatrix} 0 & 1 \\ 2 & 3 \end{vmatrix} -2 \begin{vmatrix} 3 & 1 \\ 8 & 3 \end{vmatrix} + 1 \begin{vmatrix} 3 & 0 \\ 8 & 2 \end{vmatrix}$$

$$= 6(-2) -2(9-8) + 1(6)$$

$$= -8$$

We obtain the same answer if we choose to expand the minors along the second row as follows:

$$\det \begin{vmatrix} 6 & 2 & 1 \\ 3 & 0 & 1 \\ 8 & 2 & 3 \end{vmatrix} = a_{21}\alpha_{21} + a_{22}\alpha_{22} + a_{23}\alpha_{23}$$

$$= 3(-1)^{2+1}\ |M_{21}| + (0)(-1)^{2+2}\ |M_{22}| + 1(-1)^{2+3}\ |M_{23}|$$

$$= 3(-1)\begin{vmatrix} 2 & 1 \\ 2 & 3 \end{vmatrix} + (0)\ |M_{22}| + (-1)\begin{vmatrix} 6 & 2 \\ 8 & 2 \end{vmatrix}$$

$$= -3(4) - 1(-4)$$

$$= -8$$

We usually choose to expand along the first row but another row might be easier if zeros in the row eliminate several terms.

Complete the following expansion of a fourth-order determinant.

$$\det \mathbf{A} = \begin{vmatrix} a_{11} & a_{12} & a_{13} & a_{14} \\ a_{21} & a_{22} & a_{23} & a_{24} \\ a_{31} & a_{32} & a_{33} & a_{34} \\ a_{41} & a_{42} & a_{43} & a_{44} \end{vmatrix}$$

$$= a_{11}\alpha_{11} + a_{12}\alpha_{12} + \underline{\qquad\qquad\qquad\qquad}$$

$$= a_{11}\ |M_{11}| - a_{12}\ |M_{12}| + \underline{\qquad\qquad\qquad\qquad}$$

$$= a_{11}\alpha_{11} + a_{12}\alpha_{12} + \underline{a_{13}\alpha_{13} + a_{14}\alpha_{14}}$$

$$= a_{11}\,|M_{11}| - a_{12}\,|M_{12}| + \underline{a_{13}\,|M_{13}| - a_{14}\,|M_{14}|}$$

Since we are expanding along the first row we follow the procedure given in equation (4-4) in Frame 4-13. There is a minus sign associated with $|M_{14}|$ since

$$\alpha_{14} = (-1)^{1+4}\,|M_{14}|$$

4-15

Evaluate the det **A** when

$$\mathbf{A} = \begin{bmatrix} 6 & 0 & 0 & 2 \\ 1 & 3 & 0 & 2 \\ 0 & 1 & 0 & 1 \\ 1 & 2 & 3 & 0 \end{bmatrix}$$

$$\begin{vmatrix} 6 & 0 & 0 & 2 \\ 1 & 3 & 0 & 2 \\ 0 & 1 & 0 & 1 \\ 1 & 2 & 3 & 0 \end{vmatrix} = 6|M_{11}| - 2\,|M_{14}|$$

by evaluating along the first row. However we must evaluate the third order minors $|M_{11}|$ and $|M_{14}|$ in turn

$$|M_{11}| = \begin{vmatrix} 3 & 0 & 2 \\ 1 & 0 & 1 \\ 2 & 3 & 0 \end{vmatrix} = 3 \begin{vmatrix} 0 & 1 \\ 3 & 0 \end{vmatrix} + 2 \begin{vmatrix} 1 & 0 \\ 2 & 3 \end{vmatrix} = 3(-3) + 2(3)$$

when we evaluate the minor along its first row.

$$|M_{14}| = \begin{vmatrix} 1 & 3 & 0 \\ 0 & 1 & 0 \\ 1 & 2 & 3 \end{vmatrix} = 1(-1)^{2+2} \begin{vmatrix} 1 & 0 \\ 1 & 3 \end{vmatrix} = 3$$

when we evaluate $|M_{14}|$ along its second row. Therefore det **A** = 6(−3) −2(3) = −24.

We note that the expansion of a fourth order determinant resulted in third order minors, which are determinants themselves and had to be evaluated by expanding them properly into second order minors.

If you had any difficulty in evaluating the det **A**, Frame 4-16 will give you a good review and additional practice; otherwise, skip to Frame 4-17.

4-16

For review purposes let us find the determinant of **A** when **A** is a fourth-order matrix. We want to determine the det **A** when

$$\det \mathbf{A} = \begin{vmatrix} 3 & 4 & 2 & 1 \\ 2 & 1 & 0 & 8 \\ 0 & 3 & 0 & 2 \\ 5 & 1 & 2 & 0 \end{vmatrix}$$

Examining the determinant we note that it is easiest to expand along the third row since it possesses two zeros. Then, since $\alpha_{31} = \alpha_{33} = 0$ we have

$$\det \mathbf{A} = \alpha_{32}\alpha_{32} + \alpha_{34}\alpha_{34}\cdot$$

$$= 3(-1)^{3+2}\,|M_{32}| + 2(-1)^{3+4}\,|M_{34}|$$

Evaluating $|M_{32}|$ requires deleting the third row and the second column so that

$$|M_{32}| = \begin{vmatrix} 3 & 2 & 1 \\ 2 & 0 & 8 \\ 5 & 2 & 0 \end{vmatrix} = 2(-1)^3 \begin{vmatrix} 2 & 1 \\ 2 & 0 \end{vmatrix} + 8(-1)^5 \begin{vmatrix} 3 & 2 \\ 5 & 2 \end{vmatrix}$$

$$= -2(-2) - 8(-4) = 36$$

when we expand the minor along its second row.

Similarly, expanding $|M_{34}|$ along its second row, complete the following

$$|M_{34}| = \begin{vmatrix} 3 & 4 & \\ 2 & & 0 \\ 5 & 1 & \end{vmatrix} = 2(-1)^{?} \begin{vmatrix} 4 & \\ & \end{vmatrix} + \quad (-1)^4 \begin{vmatrix} 3 & \\ & \end{vmatrix}$$

$$= \underline{\hspace{6cm}}$$

Expanding $|M_{34}|$ along its second row we have

$$|M_{34}| = \begin{vmatrix} 3 & 4 & 2 \\ 2 & 1 & 0 \\ 5 & 1 & 2 \end{vmatrix} = 2(-1)^3 \begin{vmatrix} 4 & 2 \\ 1 & 2 \end{vmatrix} + 1(-1)^4 \begin{vmatrix} 3 & 2 \\ 5 & 2 \end{vmatrix}$$

$$= -2(6) + 1(-4) = -16$$

Finally, we can evaluate the determinant as

$$\det \mathbf{A} = -3(36) - 2(-16) = -76$$

4-17

Clearly, the expansion of higher order determinants becomes a tedious process and we are fortunate indeed that we can utilize digital computers to evaluate higher order determinants.

The determinant of the square matrix **A** has several properties of interest to us. Given **A** and the transpose of **A**, which is written as **A′**, we find that

$$\det \mathbf{A} = \det \mathbf{A}'$$

In addition, for two matrices **A** and **B** that are both square matrices of order n we have

$$\det \mathbf{AB} = \det \mathbf{A} \det \mathbf{B}$$

There are several conditions that lead to a determinant equal to zero. One important condition that results in a determinant equal to zero occurs when two rows (or columns) of the determinant are proportional. This means that the corresponding elements of two rows are proportionally related by the same factor. For example, consider

$$\det \mathbf{A} = \begin{vmatrix} 4 & 1 \\ 8 & 2 \end{vmatrix} = 4(2) - 1(8) = 0$$

where the elements of the second row are twice the corresponding elements of the first row.

Verify that det $\mathbf{A}' = 0$ in this case.

$$\det \mathbf{A}' =$$

You are correct if you found that

$$\det \mathbf{A}' = \begin{vmatrix} 4 & 8 \\ 1 & 2 \end{vmatrix} = 4(2)-8 = 0$$

where in this case the elements of the first row are a factor of 4 greater than the elements of the second row.

This is a convenient stopping place in Chapter 4.

CRAMER'S RULE

4-18

One method for solving simultaneous equations by the use of determinants is called *Cramer's rule.* The solution of the nonhomogeneous matrix equation.

$$\mathbf{Ax} = \mathbf{b}$$

can be obtained, providing that the det $\mathbf{A}$ does not equal zero. Let us denote by $\mathbf{A}_i$, ($i = 1, 2, ..., n$) the matrix obtained from $\mathbf{A}$ when we replace its ith column with the column of constants $\mathbf{b}$. Then if det $\mathbf{A} \neq 0$

$$\mathbf{Ax} = \mathbf{b}$$

has the unique solution

$$x_1 = \frac{\det \mathbf{A}_1}{\det \mathbf{A}}, \qquad x_2 = \frac{\det \mathbf{A}_2}{\det \mathbf{A}}, \qquad \ldots, x_n = \frac{\det \mathbf{A}_n}{\det \mathbf{A}}$$

For example, let us obtain the solution for x_1 and x_2 when

$$\begin{bmatrix} 3 & 2 \\ 5 & 5 \end{bmatrix} \begin{bmatrix} x_1 \\ x_2 \end{bmatrix} = \begin{bmatrix} 2 \\ 1 \end{bmatrix}$$

$$x_1 = \frac{\det \mathbf{A}_1}{\det \mathbf{A}} \quad \text{where } \mathbf{A}_1 = \begin{bmatrix} 2 & 2 \\ 1 & 5 \end{bmatrix}$$

which is obtained by replacing the first column of **A** and **b**.

Evaluate x_1.

$$x_1 = \frac{\det \mathbf{A}_1}{\det \mathbf{A}} = \frac{8}{5}$$

since

$$\det \mathbf{A} = \begin{vmatrix} 3 & 2 \\ 5 & 5 \end{vmatrix} = 15 - 10 = 5$$

and

$$\det \mathbf{A}_1 = \begin{vmatrix} 2 & 2 \\ 1 & 5 \end{vmatrix} = 10 - 2 = 8$$

4-19

In a similar manner evaluate x_2.

$$x_2 = \frac{\det \mathbf{A}_2}{\det \mathbf{A}} = \frac{\begin{vmatrix} 3 & 2 \\ 5 & 1 \end{vmatrix}}{5} = \frac{-7}{5}$$

4-20

Using determinants and Cramer's rule solve the following set of equations for x_2.

$$3x_1 + 2x_2 = 1$$

$$4x_1 + 2x_3 = 2$$

$$5x_2 + 6x_3 = 0$$

$x_2 =$

You are correct if you found that $x_2 = \frac{-2}{13}$. If you are correct skip the following computations.

We have from the equations

$$\mathbf{Ax} = \mathbf{b}$$

where

$$\begin{bmatrix} 3 & 2 & 0 \\ 4 & 0 & 2 \\ 0 & 5 & 6 \end{bmatrix} \begin{bmatrix} x_1 \\ x_2 \\ x_3 \end{bmatrix} = \begin{bmatrix} 1 \\ 2 \\ 0 \end{bmatrix}$$

and

$$\begin{vmatrix} 3 & 2 & 0 \\ 4 & 0 & 2 \\ 0 & 5 & 6 \end{vmatrix} = 3 \begin{vmatrix} 0 & 2 \\ 5 & 6 \end{vmatrix} + 2(-1)^3 \begin{vmatrix} 4 & 2 \\ 0 & 6 \end{vmatrix} = 3(-10) - 2(24) = -78$$

Since $x_2 = \frac{\det \mathbf{A}_2}{\det \mathbf{A}}$ we evaluate $\det \mathbf{A}_2 = \begin{vmatrix} 3 & 1 & 0 \\ 4 & 2 & 2 \\ 0 & 0 & 6 \end{vmatrix}$ by expanding along the third row to obtain

$$\det \mathbf{A}_2 = 6(-1)^{3+3} \begin{vmatrix} 3 & 1 \\ 4 & 2 \end{vmatrix} = 6(2) = 12$$

Therefore, we find that

$$x_2 = \frac{\det \mathbf{A}_2}{\det \mathbf{A}} = \frac{-12}{78} = \frac{-2}{13}$$

Cramer's Rule is useful for hand calculations with lower order determinants. When the number of equations is large, other methods of solution are more desirable, usually employing the digital computer.

ELIMINATION METHOD

4-21

An important method for solving simultaneous equations is the *method of elimination,* often called the *Gaussian elimination method.* This method is suitable for evaluation by the digital computer and therefore is widely used. The unknown variables x_i are solved for and successively eliminated from the equations.

Let us consider the simple example solved on page 83 by the use of Cramer's rule. The simultaneous equations are

(4-5) $$3x_1 + 2x_2 = 2$$

(4-6) $$5x_1 + 5x_2 = 1$$

In order to eliminate x_1, we solve for x_1 in the equation, (4-5) obtaining

(4-7) $$x_1 = -\frac{2}{3} x_2 + \frac{2}{3}$$

Substituting into equation (4-6), we have

$$5(-\frac{2}{3}x_2 + \frac{2}{3}) + 5x_2 = 1$$

Solve for x_2.

$$x_2 = -\frac{7}{5}$$

since

$$-\frac{10}{3}x_2 + \frac{10}{3} + 5x_2 = 1$$

leads to

$$\frac{5}{3}x_2 = -\frac{7}{3}$$

Then, substituting $x_2 = -\frac{7}{5}$ into equation (4-7), we obtain

$$x_1 = -\frac{2}{3}(-\frac{7}{5}) + \frac{2}{3}$$

$$= \frac{8}{5}$$

Thus the method of elimination can be summarized as follows:

1. Solve for x_1 in the first equation in terms of the other unknowns.

2. Substitute x_1, obtained in step 1, into the remaining $(n - 1)$ equations.

3. Solve for x_2 from the first of the remaining equations.

4. Substitute x_2 into the remaining $(n - 2)$ equations.

5. Repeat the elimination process until one equation remains in one variable, x_n.

6. Solve for x_n.

7. Substitute x_n into the equation immediately preceding, which was obtained in terms of x_n and x_{n-1}.

8. Solve for x_{n-1}.

9. Repeat successively for the n unknowns.

4-22

Let us reconsider the example solved in Frame 4-20 using Cramer's rule.

(4-8) $$3x_1 + 2x_2 = 1$$

(4-9) $$4x_1 + 2x_3 = 2$$

(4-10) $$5x_2 + 6x_3 = 0$$

Solving for x_1 in equation (4-8), we have

(4-11) $$x_1 = -\frac{2}{3}x_2 + \frac{1}{3}$$

Substituting into equation (4-9), we have

(4-12) $$-\frac{8}{3}x_2 + \frac{4}{3} + 2x_3 = 2$$

or

(4-13) $$x_2 = \frac{3}{4}x_3 - \frac{1}{4}$$

Repeat the elimination process for equations (4-10) and (4-13) and obtain x_3.

$$x_3 = \frac{5}{39}$$

since substituting equation (4-13) into equation (4-10), we have

$$5\left(\frac{3}{4}x_3 - \frac{1}{4}\right) + 6x_3 = 0$$

or

$$\frac{15}{4}x_3 + 6x_3 = \frac{5}{4}$$

and therefore

$$x_3 = \frac{5}{39}.$$

4-23

Using equations (4-13) and (4-11) in that order, find x_2 and then x_1.

$$x_2 = -\frac{2}{13} \qquad \text{and} \qquad x_1 = \frac{17}{39}$$

If your answer were correct, skip this paragraph of explanation.

When we substitute $x_3 = \frac{5}{39}$ into equation (4-13), we obtain

$$\begin{aligned} x_2 &= \frac{3}{4}\left(\frac{5}{39}\right) - \frac{1}{4} \\ &= -\frac{2}{13} \end{aligned}$$

Similarly, substituting $x_2 = -\frac{2}{13}$ into equation (4-11), we have

$$\begin{aligned} x_1 &= -\frac{2}{3}\left(-\frac{2}{13}\right) + \frac{1}{3} \\ &= \frac{17}{39} \end{aligned}$$

SUMMARY

1. In this chapter you learned to express a set of simultaneous linear equations as a matrix equation. The set of m equations in n unknowns.

$$\begin{aligned} a_{11}x_1 + a_{12}x_2 + \dots + a_{1n}x_n &= b_1 \\ a_{21}x_1 + a_{22}x_2 + \dots + a_{2n}x_n &= b_2 \\ \vdots \qquad \vdots \qquad\quad & \vdots \quad\;\; \vdots \\ a_{m1}x_1 + a_{m2}x_2 + \dots + a_{mn}x_n &= b_m \end{aligned}$$

can be represented by

$$\mathbf{Ax} = \mathbf{b}$$

where **A** is an $m \times n$ matrix, **x** is an $n \times 1$ matrix, and **b** is a $m \times 1$ matrix. When $m = n$ we have a square **A** matrix.

2. The matrix equation $\mathbf{Ax} = \mathbf{b}$ is called a nonhomogeneous equation. The matrix equation $\mathbf{Ax} = \mathbf{0}$ is called a homogeneous equation where **0** is the null column matrix.

3. The determinant of a square matrix is equal to a number that may be evaluated by several procedures. The determinant of a second order matrix is

$$\det \mathbf{A} = \begin{vmatrix} a_{11} & a_{12} \\ a_{21} & a_{22} \end{vmatrix} = a_{11}a_{22} - a_{12}a_{21}$$

   The determinant of a third order matrix can be evaluated by a procedure using a schedule as illustrated in Frame 4-10.

4. A minor of a determinant is another determinant formed by removing an equal number of rows and columns from the original determinant. The order of the minor indicates the number of rows (or columns) in the minor. $|M_{ij}|$ refers to the minor of $|\mathbf{A}|$ obtained by removing the ith row and the jth column.

5. The cofactor, α_{ij}, of an element of a determinant is found by giving an appropriate sign to the minor obtained when the row and column corresponding to that element are removed.

$$\alpha_{ij} = (-1)^{i+j} |M_{ij}|$$

6. Cramer's rule is used for solving a set of simultaneous equations. The solution of the nonhomogeneous matrix equation

$$\mathbf{Ax} = \mathbf{b}$$

   is unique when $\det \mathbf{A} \neq 0$
   and is

$$x_i = \frac{\det \mathbf{A}_i}{\det \mathbf{A}}$$

where $\mathbf{A}_i$ is the matrix obtained from **A** when we replace its ith column with the column of constants **b**.

7. The method of elimination is used to solve simultaneous equations and is particularly suitable for digital computer calculations. Starting with the first equation, the unknown variable, x_1, is solved for and substituted into the remaining equations in order to eliminate the variable x_1. Following in this manner, all but one of the unknown variables are eliminated from the equations. When x_n is obtained, then it is used to obtain x_{n-1} and successive variables.

EXERCISES

1. Can you obtain the det **A** when

$$\mathbf{A} = \begin{bmatrix} 3 & 2 & 1 \\ & & \\ 4 & 6 & 2 \end{bmatrix} ?$$

det **A** =

2. Obtain the matrix equation representing

$$4x_1 + 6x_2 = 7$$

$$5x_1 + 7x_3 = 8$$

$$2x_2 + 4x_3 = 0$$

$$\begin{bmatrix} \quad & \quad & \quad \\ & & \\ & & \end{bmatrix} \begin{bmatrix} \quad \\ \\ \\ \end{bmatrix} = \begin{bmatrix} \quad \\ \\ \\ \end{bmatrix}$$

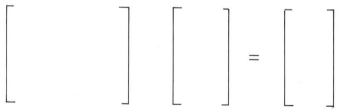

3. Obtain the minor $|M_{22}|$ of the determinant **P**

$$\mathbf{P} = \begin{vmatrix} p_{11} & p_{12} & p_{13} \\ p_{21} & p_{22} & p_{23} \\ p_{31} & p_{32} & p_{33} \end{vmatrix} \qquad |M_{22}| =$$

$|M_{22}|$ is a ______________ order determinant.

4. Evaluate the determinant of **R** when

$$\mathbf{R} = \begin{bmatrix} 5 & 0 & 2 \\ 1 & 6 & 2 \\ 3 & 0 & 1 \end{bmatrix}$$

5. Expand the determinant of the matrix **Q** by minors.

$$\mathbf{Q} = \begin{bmatrix} 2 & 1 & 0 & 5 \\ 3 & 1 & 5 & 0 \\ 2 & 0 & 1 & 0 \\ 8 & 0 & 2 & 2 \end{bmatrix}$$

6. Use Cramer's rule to solve for x_3 in the following set of simultaneous equations.

$$2x_1 + 6x_2 = 0$$

$$4x_1 + 2x_3 = 2$$

$$x_2 + x_3 = 2$$

7. Given a matrix **P**, det **P′** = ______________________________

8. det **PQ** = ____________ where **P** and **Q** are both square matrices of order *n*.

9.

$$\det \mathbf{A} = \begin{vmatrix} 6 & 2 & 1 \\ 3 & 0 & 4 \\ 18 & 6 & 3 \end{vmatrix} = 0$$

By inspection of the det **A**, state why the det **A** = 0.

10. The matrix equation **Qy** = **c**, where **c** is a column vector of constants, is called a ____________________ equation.

11. Use the method of elimination to solve for the unknowns x_1, x_2, and x_3 when

$$2x_1 - x_2 - x_3 = 1$$

$$3x_1 + x_2 + x_3 = 4$$

$$2x_1 + 2x_2 - 2x_3 = 8$$

The answers to these exercises appear on page 238.

5 the rank, trace, and adjoint of a matrix

5-1

Rank is defined as the order of the largest nonzero determinant that can be obtained from the elements of a matrix. This definition applies to both square and rectangular (nonsquare) matrices. Therefore, a nonzero matrix **A** is said to have rank r if at least one of its r-square minors is different from zero, while every $(r + 1)$ or larger square minor, if any, is equal to zero. The rank of the **A** matrix can be found by starting with the largest determinants of order m and evaluating them to ascertain if one of them is nonzero. If so, then the rank of the matrix is equal to m. If all the determinants of order m are equal to zero then we start evaluating the determinants of $(m - 1)$. Continuing in this manner we eventually find the rank r of the matrix, which is the order of the largest nonzero determinant.

Let us find the rank of the matrix

$$\mathbf{A} = \begin{bmatrix} 6 & 2 \\ 3 & 1 \end{bmatrix}$$

The det $\mathbf{A} = \begin{vmatrix} 6 & 2 \\ 3 & 1 \end{vmatrix} = 0$ since the det **A** is the minor of largest order for **A**.

Is the rank of **A** equal to 0 or 1?

The rank of **A** is equal to 1.
This is true since the order of the largest nonzero minor of **A** is equal to 1. Actually, there are four nonzero minors of order 1 that can be selected, for example $|M_{11}| = 6$.

5-2

Find the rank of matrix **B** where

$$\mathbf{B} = \begin{bmatrix} 0 & 3 \\ 0 & 1 \end{bmatrix}$$

The rank of **B** is equal to ________________.

The rank of **B** is equal to 1.

Explanation:

$$\det \mathbf{A} = \begin{vmatrix} 0 & 3 \\ 0 & 1 \end{vmatrix} = 0$$

There are two nonzero minors and, therefore, the rank of the matrix is 1.

$$|M_{12}| = 3, \quad |M_{22}| = 1.$$

5-3

The rank of the matrix

$$\mathbf{C} = \begin{bmatrix} 1 & 2 & 3 \\ 2 & 3 & 4 \\ 3 & 5 & 7 \end{bmatrix}$$

is equal to ____________.

The rank of the matrix **C** is equal to *2.*

Explanation:

Since the matrix **C** is square, the largest minor is the determinant of order 3. However, det **C** = 0. Then, trying a minor of order 2, we find that

$$|M_{11}| = \begin{vmatrix} 3 & 4 \\ 5 & 7 \end{vmatrix}$$

and the rank equals 2.

5-4

A square matrix **A** of order n is called *nonsingular* if its rank is equal to n $(r = n)$, and therefore, det **A** $\neq 0$. Otherwise **A** is *singular* and det **A** $= 0$.

For example, the matrix considered in Frame 5-3 is singular since

$$\det \mathbf{C} = \begin{vmatrix} 1 & 2 & 3 \\ 2 & 3 & 4 \\ 3 & 5 & 7 \end{vmatrix} = 0$$

Consider the following matrix.

$$\mathbf{D} = \begin{bmatrix} 8 & 0 & 0 \\ 0 & 2 & 0 \\ 0 & 0 & 4 \end{bmatrix}$$

The rank of the diagonal matrix **D** is equal to ________________ and the matrix is ________________________.
(singular/nonsingular)

The rank of the diagonal matrix **D** is equal to 3 and the matrix is nonsingular.

If your answer was correct proceed to Frame 5-5; otherwise read the following explanatory paragraph.

The determinant of **D** is nonzero and is

$$\det \mathbf{D} = \begin{vmatrix} 8 & 0 & 0 \\ 0 & 2 & 0 \\ 0 & 0 & 4 \end{vmatrix} = 8(2)4 = 64$$

Therefore the rank of the matrix is equal to 3. Since the determinant of **D** is not equal to zero we label it as nonsingular.

5-5

The definition of the rank of a matrix applies to rectangular matrices as well as to square ones. In the case of a rectangular matrix we determine the order of the largest nonzero minor of the matrix. Recall that a minor is a determinant and is square, that is, $m \times m$.

As an illustration, consider the 3×4 rectangular matrix

$$\mathbf{F} = \begin{bmatrix} 5 & 6 & 2 & 1 \\ 0 & 0 & 8 & 2 \\ 0 & 0 & 4 & 2 \end{bmatrix}$$

One 3×3 determinant formed from the first three columns is

$$\begin{vmatrix} 5 & 6 & 2 \\ 0 & 0 & 8 \\ 0 & 0 & 4 \end{vmatrix} = 5 \begin{vmatrix} 0 & 8 \\ 0 & 4 \end{vmatrix} = 0$$

Since the foregoing determinant is zero, we can select another 3 X 3 determinant from the second, third, and fourth columns, to yield

$$\begin{vmatrix} 6 & 2 & 1 \\ 0 & 8 & 2 \\ 0 & 4 & 2 \end{vmatrix} = 6 \begin{vmatrix} 8 & 2 \\ 4 & 2 \end{vmatrix} = 6(16 - 8) = 48$$

Therefore the rank of the **F** matrix is 3.

Verify that the rank is 3 by evaluating the determinant formed from the first, third, and fourth columns.

$$\begin{vmatrix} & & \\ & & \\ & & \end{vmatrix} =$$

$$\begin{vmatrix} 5 & 2 & 1 \\ 0 & 8 & 2 \\ 0 & 4 & 2 \end{vmatrix} = 5(8) = 40$$

and therefore the matrix is of rank 3.

5-6

Quite often we are interested in solving a set of linear algebraic equations that we can express in matrix form, as we found in Chapter 4. A set of simultaneous linear equations is said to be *consistent* if there is a solution, and *inconsistent* if there is no solution. We can determine if the equations are consistent by comparing the rank of the coefficient matrix **A** with the augmented matrix. Consider a linear matrix equation representing a set of n linear algebraic equations

$$\mathbf{A}\mathbf{x} = \mathbf{b}$$

The *augmented matrix* is the matrix obtained by augmenting the matrix **A** with the column **b** to obtain the matrix

$$\mathbf{A}^b = \begin{bmatrix} a_{11} & a_{12} & \cdots & a_{1n} & b_1 \\ a_{21} & a_{22} & \cdots & a_{2n} & b_2 \\ \cdot & \cdot & & \cdot & \cdot \\ \cdot & \cdot & & \cdot & \cdot \\ \cdot & \cdot & & \cdot & \cdot \\ a_{n1} & a_{n2} & \cdots & a_{nn} & b_n \end{bmatrix}$$

which has n rows and $n + 1$ columns.

Write the augmented matrix for a set of three simultaneous nonhomogeneous linear equations

$$\mathbf{A}^b =$$

$$\mathbf{A}^b = \begin{bmatrix} a_{11} & a_{12} & a_{13} & b_1 \\ a_{21} & a_{22} & a_{23} & b_2 \\ a_{31} & a_{32} & a_{33} & b_3 \end{bmatrix}$$

5-7

A set of n simultaneous equations in n unknowns is consistent if the rank of the coefficient matrix **A** is equal to the rank of the augmented matrix $\mathbf{A}^b$. If the rank of **A** is less than the rank of $\mathbf{A}^b$ the equations are inconsistent and thus there is no solution.

Consider the following matrix equation

$$\mathbf{Ax} = \mathbf{b}$$

$$\begin{bmatrix} 1 & 3 \\ 2 & 1 \end{bmatrix} \begin{bmatrix} x_1 \\ x_2 \end{bmatrix} = \begin{bmatrix} 3 \\ 0 \end{bmatrix}$$

The rank of **A** is equal to ________________

The augmented matrix is

$$\mathbf{A}^b =$$

The rank of the augmented matrix is ________________

The set of equations is ________________ (consistent/inconsistent) and there is ____ (no/a) solution.

The rank of **A** is equal to 2 since det $\mathbf{A} \neq 0$. The augmented matrix is

$$\mathbf{A}^b = \begin{bmatrix} 1 & 3 & 3 \\ 2 & 1 & 0 \end{bmatrix}$$

The rank of the augmented matrix is 2 since the largest determinant of $\mathbf{A}^b$ is of order 2.

The set of equations is *consistent* and there is a solution.

5-8

There have been quite a few definitions on the last few pages so let's review them with a few questions. Please fill in the blanks.

1. A set of simultaneous linear equations is consistent if there is a ____________________.

2. The augmented matrix is obtained by augmenting the matrix **A** with the column matrix ____________.

3. A set of n equations in n unknowns is consistent if the rank of **A** is equal to ____________________.

1. A set of simultaneous linear equations is consistent if there is a solution.

2. The augmented matrix is obtained by augmenting the matrix **A** with the column matrix **b**.

3. A set of n equations in n unknowns is consistent if the rank of **A** is equal to the rank of $\mathbf{A}^b$.

5-9

When the equations are consistent for a set of n equations in n unknowns and the matrices **A** and $\mathbf{A}^b$ each have rank n (that is, the matrix is nonsingular and det $\mathbf{A} \neq 0$, then there is a *unique* solution for the n equations. If the rank of each matrix is m and $m < n$, then the matrix is singular and there is an infinite number of solutions to the equations. Specifically, there will be $n - m$ unknowns that can be given arbitrary values while the remaining m unknowns will be related uniquely to them.

Consider the matrix equation representing a set of two linear equations:

$$\begin{bmatrix} 3 & 2 \\ 1 & 4 \end{bmatrix} \begin{bmatrix} x_1 \\ x_2 \end{bmatrix} = \begin{bmatrix} 1 \\ 1 \end{bmatrix}$$

The solution to this set of two equations is ______________
unique/nonunique

The solution to this set of equations is unique. If you have questions, read on; otherwise proceed to Frame 5-10.

The det **A** is nonsingular, and since **A** is of order 2 we find that **A** is of rank 2.

$$\det \mathbf{A} = \begin{vmatrix} 3 & 2 \\ 1 & 4 \end{vmatrix} = 10$$

The augmented matrix is

$$\mathbf{A}^b = \begin{bmatrix} 3 & 2 & 1 \\ 1 & 4 & 1 \end{bmatrix}$$

The largest nonzero determinant (which must be square) is of order 2 and therefore $\mathbf{A}^b$ is of rank 2. Since there are two equations to be solved ($n = 2$) and both the rank of **A** and $\mathbf{A}^b$ equal 2 we can say that the solution of the set of equations $\mathbf{Ax} = \mathbf{b}$ is unique.

5-10

Consider the set of equations

$$2x_1 + 3x_2 = 1$$

$$-4x_1 - 6x_2 = -2$$

1. The rank of the matrix **A** is __________________ .

2. The rank of the augmented matrix $\mathbf{A}^b$ is __________________ .

3. The equations are __________________ .
   consistent/inconsistent

4. The solution to this set of equations is __________________ .
   unique/nonunique

$$\begin{bmatrix} 2 & 3 \\ -4 & -6 \end{bmatrix} \begin{bmatrix} x_1 \\ x_2 \end{bmatrix} = \begin{bmatrix} 1 \\ -2 \end{bmatrix}$$

$$\det \mathbf{A} = \begin{vmatrix} 2 & 3 \\ -4 & -6 \end{vmatrix} = 0$$

1. The rank of **A** is 1.

$$\mathbf{A}^b = \begin{bmatrix} 2 & 3 & 1 \\ -4 & -6 & -2 \end{bmatrix}$$

All the determinants of order 2 formed from $\mathbf{A}^b$ are equal to zero. Therefore,

2. the rank of $\mathbf{A}^b$ is 1. The ranks of **A** and $\mathbf{A}^b$ are the same, the equations
3. are consistent and have a solution. The solution of this set of equations
4. is nonunique since the rank $r = 1$ is less than the number of unknowns $n = 2$.

5-11

Determine how x_1 is related to x_2 by solving for x_1 in either of the equations.

Solving the first equation, we obtain

$$2x_1 = 1 - 3x_2$$

$$\text{or } x_1 = -\frac{3}{2}x_2 + \frac{1}{2}$$

Therefore x_2 can be assigned any arbitrary value and then x_1 is determined uniquely. In matrix form

$$\mathbf{x} = \begin{bmatrix} x_1 \\ x_2 \end{bmatrix} = \begin{bmatrix} (-\frac{3}{2}x_2 + \frac{1}{2} \\ x_2 \end{bmatrix} = \begin{bmatrix} -3/2 \\ 1 \end{bmatrix} x_2 + \begin{bmatrix} 1/2 \\ 0 \end{bmatrix}$$

where x_2 is arbitrarily assigned.

5-12

Now let us briefly review the conditions for the solution of n simultaneous equations. A set of equations is consistent and have a solution when the rank of $\mathbf{A}$, $r(\mathbf{A})$, is equal to the rank of ____________. The solution of a set of n consistent equations is unique when the rank of $\mathbf{A}$, $r(\mathbf{A})$ equals the rank of $\mathbf{A}^b$, $r(\mathbf{A}^b)$, and the rank equals __________ .

A set of equations are consistent and have a solution when the rank of $\mathbf{A}$ is equal to the rank of $\mathbf{A}^b$.

The solution of a set of n consistent equations is unique when the rank of $\mathbf{A}$ equals the rank of $\mathbf{A}^b$ and the rank equals n.

5-13

The solution of a set of n consistent equations is nonunique when $r(\mathbf{A}) = r(\mathbf{A}^b) = m$ and $m < n$. There are ________________ arbitrarily assigned unknowns which then assign unique values to the ____________ remaining unknowns.

There are $n - m$ arbitrarily assigned unknowns, which then assign unique values to the m remaining unknowns.

This may be convenient stopping place for you.

THE TRACE OF A MATRIX

5-14

The sum of the elements in the major (or principal) diagonal of any square matrix **A** is called the trace of **A** and is written as $tr(\mathbf{A})$. Therefore, considering the general matrix **A** where

$$\mathbf{A} = \begin{bmatrix} a_{11} & a_{12} & \cdots & a_{1n} \\ a_{21} & a_{22} & \cdots & a_{2n} \\ \cdot & \cdot & & \\ \cdot & \cdot & & \\ \cdot & \cdot & & \\ a_{n1} & a_{n2} & \cdots & a_{nn} \end{bmatrix}$$

we obtain

$$tr(\mathbf{A}) = a_{11} + a_{22} + \ldots + a_{nn} = \sum_{i=1}^{n} a_{ii}$$

Evaluate the trace of the matrix

$$\mathbf{E} = \begin{bmatrix} 4 & 2 & 1 \\ 0 & 3 & 4 \\ 6 & 0 & -2 \end{bmatrix}$$

$tr(\mathbf{E}) =$ ______________

$$tr(\mathbf{E}) = 4 + 3 - 2 = 5$$

COFACTOR MATRIX

5-15

In order to proceed toward the development of an inverse matrix operation, let us define a *cofactor matrix.* A cofactor matrix, $\mathbf{A}^C$, is obtained by replacing each element of a square matrix $\mathbf{A}$ by its corresponding cofactor α_{ij}. Thus, we replace each element a_{ij} by α_{ij} where α_{ij} is determined as discussed in Chapter 4 (see page 73) and is

$$\alpha_{ij} = (-1)^{(i+j)} \quad |M_{ij}|$$

where $|M_{ij}|$ is the minor corresponding to the removal of the ith row and jth column.

Find the cofactor matrix for

$$\mathbf{B} = \begin{bmatrix} 3 & 2 \\ 4 & -1 \end{bmatrix}, \quad \mathbf{B}^C = \begin{bmatrix} & \\ & \end{bmatrix}$$

$$\mathbf{B}^c = \begin{bmatrix} -1 & -4 \\ -2 & 3 \end{bmatrix}$$

noting that

$$\alpha_{12} = (-1)^3\ 4 = -4$$

and

$$\alpha_{21} = (-1)^3\ 2 = -2$$

5-16

Complete the cofactor matrix $\mathbf{A}^c$ when

$$\mathbf{A} = \begin{bmatrix} 1 & 2 & 3 \\ 2 & 3 & 2 \\ 3 & 3 & 4 \end{bmatrix}$$

$$\mathbf{A}^c = \begin{bmatrix} 6 & \_\_ & -3 \\ 1 & \_\_ & 3 \\ \_\_ & 4 & \_\_ \end{bmatrix}$$

$$\mathbf{A}^c = \begin{bmatrix} 6 & -2 & -3 \\ 1 & -5 & 3 \\ -5 & 4 & -1 \end{bmatrix}$$

If you answered correctly, proceed to Frame 5-17. If not, please read the following first:

$$\mathbf{A}^c = \begin{bmatrix} \alpha_{11} & \alpha_{12} & \alpha_{13} \\ \alpha_{21} & \alpha_{22} & \alpha_{23} \\ \alpha_{31} & \alpha_{32} & \alpha_{33} \end{bmatrix}$$

$$\alpha_{12} = (-1)^3 \begin{vmatrix} 2 & 2 \\ 3 & 4 \end{vmatrix} = -1(8-6) = -2$$

$$\alpha_{22} = (-1)^4 \begin{vmatrix} 1 & 3 \\ 3 & 4 \end{vmatrix} = (4-9) = -5$$

$$\alpha_{31} = (-1)^4 \begin{vmatrix} 2 & 3 \\ 3 & 2 \end{vmatrix} = (4-9) = -5$$

$$\alpha_{33} = (-1)^6 \begin{vmatrix} 1 & 2 \\ 2 & 3 \end{vmatrix} = (3-4) = -1$$

5-17

If we have a square matrix **A** and its cofactor matrix $\mathbf{A}^c$ where

$$\mathbf{A}^c = \begin{bmatrix} \alpha_{11} & \alpha_{12} & \cdots & \alpha_{1n} \\ \alpha_{21} & \alpha_{22} & \cdots & \alpha_{2n} \\ \cdot & \cdot & & \cdot \\ \cdot & \cdot & & \cdot \\ \cdot & \cdot & & \cdot \\ \alpha_{n1} & \alpha_{n2} & \cdots & \alpha_{nn} \end{bmatrix}$$

then we define the *adjoint matrix* of **A** as the transpose of the cofactor matrix so that

$$adj\ \mathbf{A} = \mathbf{A}^{c'}$$

Therefore,

$$adj\ \mathbf{A} = \begin{bmatrix} \alpha_{11} & \alpha_{21} & \cdots & \alpha_{n1} \\ \alpha_{12} & \alpha_{22} & \cdots & \alpha_{n2} \\ \cdot & \cdot & & \cdot \\ \cdot & \cdot & & \cdot \\ \cdot & \cdot & & \cdot \\ \alpha_{1n} & \alpha_{2n} & \cdots & \alpha_{nn} \end{bmatrix}$$

5-18

For a matrix

$$\mathbf{A} = \begin{bmatrix} 3 & 2 \\ -1 & 4 \end{bmatrix}$$

complete the following adjoint matrix of **A**

$$adj\ \mathbf{A} = \begin{bmatrix} 4 & \_\_\_ \\ \_\_\_ & \_\_\_ \end{bmatrix}$$

$$adj\ \mathbf{A} = \begin{bmatrix} 4 & -2 \\ 1 & 3 \end{bmatrix}$$

For example,

$$\alpha_{12} = (-1)^3\,(-1) = 1$$

which then is transposed to the position in the second row and first column.

5-19

Consider the matrix **A** in Frame 5-16. We found that when

$$\mathbf{A} = \begin{bmatrix} 1 & 2 & 3 \\ 2 & 3 & 2 \\ 3 & 3 & 4 \end{bmatrix}$$

then

$$\mathbf{A}^C = \begin{bmatrix} 6 & -2 & -3 \\ 1 & -5 & 3 \\ -5 & 4 & 1 \end{bmatrix}$$

Complete the adjoint of **A**

$$adj\ \mathbf{A} = \begin{bmatrix} 6 & \_\_\_ & -5 \\ -2 & \_\_\_ & \_\_\_ \\ \_\_\_ & 3 & 1 \end{bmatrix}$$

$$adj\ \mathbf{A} = \begin{bmatrix} 6 & 1 & -5 \\ -2 & -5 & 4 \\ -3 & 3 & 1 \end{bmatrix}$$

since $adj\ \mathbf{A} = \mathbf{A}^{c'}$.

5-20

The adjoint matrix possesses a very important multiplication property. When we multiply the matrix **A** by its adjoint we have

$$\mathbf{A}\cdot(adj\ \mathbf{A}) = \begin{bmatrix} a_{11} & a_{12} & \cdots & a_{1n} \\ a_{21} & a_{22} & \cdots & a_{2n} \\ \cdot & \cdot & & \cdot \\ \cdot & \cdot & & \cdot \\ \cdot & \cdot & & \cdot \\ a_{n1} & a_{n2} & \cdots & a_{nn} \end{bmatrix} \begin{bmatrix} \alpha_{11} & \alpha_{21} & \alpha_{n1} \\ \alpha_{12} & \alpha_{22} & \alpha_{n2} \\ \cdot & \cdot & \cdot \\ \cdot & \cdot & \cdot \\ \cdot & \cdot & \cdot \\ \alpha_{1n} & \alpha_{2n} & \ldots\alpha_{nn} \end{bmatrix}$$

$$= |\mathbf{A}|\ \mathbf{I}$$

where **I** is the identity matrix. The multiplication, in this case, is commutative and we can write

$$\mathbf{A}\cdot(adj\ \mathbf{A}) = (adj\ \mathbf{A})\cdot\mathbf{A} = |\mathbf{A}|\ \mathbf{I}$$

Therefore multiplying **A** by the adjoint of **A** yields a matrix that has the value of the determinant of **A** on the diagonal so that

$$\mathbf{A}\cdot(adj\ \mathbf{A}) = \begin{bmatrix} |\mathbf{A}| & 0 & \cdots & 0 \\ 0 & |\mathbf{A}| & \cdots & 0 \\ \cdot\cdot & \cdot & & \\ \cdot & \cdot & & \\ \cdot & \cdot & & \\ 0 & 0 & \cdots & |\mathbf{A}| \end{bmatrix}$$

Find the product $\mathbf{A}\cdot(adj\ \mathbf{A})$ for the example considered on page 111 and 112.

$$\mathbf{A} = \begin{bmatrix} 3 & 2 \\ -1 & 4 \end{bmatrix} \quad \text{and} \quad adj\ \mathbf{A} = \begin{bmatrix} 4 & -2 \\ 1 & 3 \end{bmatrix}$$

$\mathbf{A} \cdot adj\ \mathbf{A} =$

$$\mathbf{A} \cdot (adj\ \mathbf{A}) = \begin{bmatrix} 3 & 2 \\ -1 & 4 \end{bmatrix} \begin{bmatrix} 4 & -2 \\ 1 & 3 \end{bmatrix} = \begin{bmatrix} 14 & 0 \\ 0 & 14 \end{bmatrix}$$

Note that

$$(adj\ \mathbf{A}) \cdot \mathbf{A} = \begin{bmatrix} 4 & -2 \\ 1 & 3 \end{bmatrix} \begin{bmatrix} 3 & 2 \\ -1 & 4 \end{bmatrix} = \begin{bmatrix} 14 & 0 \\ 0 & 14 \end{bmatrix}$$

5-21

Find the det $\mathbf{A}$ and verify that

$$(adj\ \mathbf{A})\cdot\mathbf{A} = \begin{bmatrix} |\mathbf{A}| & 0 \\ 0 & |\mathbf{A}| \end{bmatrix}$$

det $\mathbf{A}$ =

$$\det \mathbf{A} = \begin{vmatrix} 3 & 2 \\ -1 & 4 \end{vmatrix} = 12 - (-2) = 14$$

so that

$$\mathbf{A} \cdot (adj\ \mathbf{A}) = \begin{bmatrix} |\mathbf{A}| & 0 \\ 0 & |\mathbf{A}| \end{bmatrix} = |\mathbf{A}|\ \mathbf{I}$$

5-22

Let us take the determinant of the relation

$$\mathbf{A} \cdot (adj\ \mathbf{A}) = (adj\ \mathbf{A}) \cdot \mathbf{A} = |\mathbf{A}|\ \mathbf{I}$$

obtaining

$$|\mathbf{A}| \cdot |adj\ \mathbf{A}| = |adj\ \mathbf{A}| \cdot |\mathbf{A}| = |\mathbf{A}|^n$$

We note that the determinant of $|\mathbf{A}|\ \mathbf{I}$ is $|\mathbf{A}|^n$ since

$$\det (|\mathbf{A}|\ \mathbf{I}) = \det \begin{bmatrix} |\mathbf{A}| & 0 & \dots & 0 \\ 0 & |\mathbf{A}| & \dots & 0 \\ \cdot & \cdot & & \cdot \\ \cdot & \cdot & & \cdot \\ \cdot & \cdot & & \cdot \\ 0 & 0 & \dots & |\mathbf{A}| \end{bmatrix} = |\mathbf{A}|^n$$

for an nth order matrix.

For the example considered on pages 114 and 115, we have

$$\mathbf{A} = \begin{bmatrix} 3 & 2 \\ -1 & 4 \end{bmatrix} \qquad adj\ \mathbf{A} = \begin{bmatrix} 4 & -2 \\ 1 & 3 \end{bmatrix}$$

Find the product of the determinants

$$|\mathbf{A}| \cdot |adj\ \mathbf{A}| =$$

$$|\mathbf{A}| \cdot |adj\ \mathbf{A}| = (14)(14) = 196$$

$$= |\mathbf{A}|^2$$

5-23

If $\mathbf{A}$ is an nth order square matrix and is nonsingular, then when

$$|\mathbf{A}| \cdot |adj\ \mathbf{A}| = |\mathbf{A}|^n$$

we can divide both sides by $|\mathbf{A}|$ to obtain

$$|adj\ \mathbf{A}| = |\mathbf{A}|^{n-1}$$

when

$$\mathbf{A} = \begin{bmatrix} b & 0 & 0 \\ 0 & b & 0 \\ 0 & 0 & b \end{bmatrix}$$

Find $adj\ \mathbf{A}$, $\mathbf{A} \cdot (adj\ \mathbf{A})$, and $|\mathbf{A}| \cdot |adj\ \mathbf{A}|$.

$$adj\ \mathbf{A} = \begin{bmatrix} b^2 & 0 & 0 \\ 0 & b^2 & 0 \\ 0 & 0 & b^2 \end{bmatrix} = b^2\mathbf{I} = \mathbf{A} \cdot \mathbf{A} = \mathbf{A}^2$$

$$\mathbf{A} \cdot (adj\ \mathbf{A}) = \begin{bmatrix} b^3 & 0 & 0 \\ 0 & b^3 & 0 \\ 0 & 0 & b^3 \end{bmatrix} = \mathbf{A}^3 = b^3\mathbf{I}$$

Since

$$|\mathbf{A}| = b^3 \text{ and } |adj\ \mathbf{A}| = b^6 \text{ we obtain}$$

$$|\mathbf{A}| \cdot |adj\ \mathbf{A}| = b^3(b^6) = b^9$$

or alternatively,

$$|\mathbf{A}| \cdot |adj\ \mathbf{A}| = |\mathbf{A}|^3 = (b^3)^3 = b^9$$

SUMMARY

1. The *rank* of a matrix is defined as the order of the largest nonzero determinant that can be obtained from the elements of a matrix.

2. A square matrix $\mathbf{A}$ of order n is called *nonsingular* if its rank is equal to n, and therefore det $\mathbf{A} \neq 0$. Otherwise $\mathbf{A}$ is *singular* and det $\mathbf{A} = 0$.

3. A set of simultaneous equations is said to be *consistent* if there is a solution to the equations, and *inconsistent* if there is no solution. A set of n simultaneous equations in n unknowns is consistent if the rank of $\mathbf{A}$ is equal to the rank of the augmented matrix $\mathbf{A}^b$ for the equations $\mathbf{Ax} = \mathbf{b}$.

4. For a set of n equations in n unknowns that are consistent and the rank of $\mathbf{A}$ and $\mathbf{A}^b$ are both equal to n, there is a *unique* solution for the n equations.

5. If the rank of **A** and $\mathbf{A}^b$ is m where $m < n$, then the matrix is singular (det $\mathbf{A} = 0$) and there will be $n - m$ unknowns that can be given arbitrary values while the remaining m unknowns will be related uniquely to them.

Equations $\mathbf{Ax} = \mathbf{b}$ and Nature of Solution

Types of Equations	Rank	Nature of Solution
n Consistent equations	$r(\mathbf{A}) = r(\mathbf{A}^b) = n$	unique solution
	$r(\mathbf{A}) = r(\mathbf{A}^b) = m < n$	infinite number of solutions with $n - m$ arbitrary unknowns
Inconsistent equations	$r(\mathbf{A}) < r(\mathbf{A}^b)$	no solution

6. The sum of the elements on the major diagonal of a matrix **A** is called the *trace* of **A** and is written as *tr* (**A**).

7. A *cofactor* matrix, $\mathbf{A}^c$, is obtained by replacing each element of a square matrix **A** by its corresponding cofactor α_{ij}.

8. The *adjoint* matrix of **A**, is the transpose of the cofactor matrix so that *adj* $\mathbf{A} = \mathbf{A}^{c'}$

9. Two properties of the adjoint matrix are

$$\mathbf{A} \cdot (adj\ \mathbf{A}) = (adj\ \mathbf{A}) \cdot \mathbf{A} = |\mathbf{A}|\ \mathbf{I}$$

and

$$|\mathbf{A}| \cdot |adj\ \mathbf{A}| = |adj\ \mathbf{A}| \cdot |\mathbf{A}| = |\mathbf{A}|^n$$

EXERCISES

1. Find the rank of the matrix **P** where

$$\mathbf{P} = \begin{bmatrix} 1 & 2 & 3 \\ -4 & 0 & 5 \end{bmatrix}$$

$r(\mathbf{P})$ = ________________

2. A set of equations is represented by $\mathbf{Ax} = \mathbf{b}$ where

$$\mathbf{A} = \begin{bmatrix} 1 & 2 & 3 \\ 1 & 2 & 5 \\ 2 & 4 & 8 \end{bmatrix} \quad \text{and } \mathbf{b} = \begin{bmatrix} 1 \\ 3 \\ 0 \end{bmatrix}$$

The matrix **A** is ________________.
singular/nonsingular

The set of equations is ________________.
consistent/inconsistent

3. A set of equations is represented by $\mathbf{Ax} = \mathbf{b}$ where

$$\mathbf{A} = \begin{bmatrix} 3 & 2 \\ 1 & 2 \end{bmatrix} \quad \text{and } \mathbf{b} = \begin{bmatrix} 1 \\ 0 \end{bmatrix}$$

There is a ________________ solution.
unique/nonunique

4. Find the trace of **A** where

$$\mathbf{A} = \begin{bmatrix} 6 & 2 & 1 \\ 3 & -1 & 6 \\ 0 & 1 & 3 \end{bmatrix}$$

tr **A** = ________________

5. Find the cofactor matrix, $\mathbf{A}^c$, when

$$\mathbf{A} = \begin{bmatrix} 2 & -1 & 4 \\ 0 & 4 & -2 \\ 1 & -3 & 2 \end{bmatrix} \qquad \mathbf{A}^c = \begin{bmatrix} \quad & \quad & \quad \\ & & \\ & & \end{bmatrix}$$

6. Find the adjoint of the **A** matrix in problem 5.

$$adj\ \mathbf{A} = \begin{bmatrix} \quad & \quad & \quad \\ & & \\ & & \end{bmatrix}$$

7. Complete the following equation for a property of the adjoint matrix by filling in the blank space

$$\mathbf{A} \cdot (adj\ \mathbf{A}) = |\mathbf{A}|\ \_\_\_\_\_\_\_\_$$

8. Complete the following equation for a property of the determinant of the adjoint matrix by filling in the blank space.

$$|\mathbf{A}| \cdot |adj\ \mathbf{A}| = \_\_\_\_\_\_\_\_\_\_$$

The answers are on page 241.

6 the inverse of a matrix

6-1

In the algebra of numbers we have the division operation so that the solution of the algebraic equation $ax = b$ is

$$x = \frac{b}{a}.$$

The matrix operation that is analogous to division is known as *matrix inversion* and a method of calculating the *inverse* of a matrix is given in this chapter. The inverse of a square matrix **A** is written as $\mathbf{A}^{-1}$ and it is defined as the matrix, which when multiplied by the original **A** matrix results in the identity matrix. Therefore,

(6-1) $$\mathbf{A}^{-1}\mathbf{A} = \mathbf{A}\mathbf{A}^{-1} = \mathbf{I}$$

and the matrix and its inverse are commutative.

In the algebra of numbers we write $ab = 1$, and therefore $b = a^{-1}$ and we say that b is the inverse (or reciprocal) of a.

In matrix algebra we have

$$\mathbf{AB} = \mathbf{I}$$

from which we infer that **B** is the ______________ of **A**; or in equation form,

B = ______________________________

In matrix algebra we have **AB** = **I** from which we infer that **B** is the *inverse* of **A**; or in equation form

(6-2) $$\mathbf{B} = \mathbf{A}^{-1}$$

When we solve a linear equation, $ax = b$, we have

$$x = \frac{b}{a}$$

An analogous matrix equation representing a set of n simultaneous linear equations in n unknown is written as

(6-3) $$\mathbf{Ax} = \mathbf{b}$$

where **A** is a square matrix of coefficients. Premultiply both sides of equation (6-3) by $\mathbf{A}^{-1}$ to obtain

$$\mathbf{A}^{-1}\mathbf{Ax} = \mathbf{A}^{-1}\mathbf{b}$$

Since equation (6-1) states that $\mathbf{A}^{-1}\mathbf{A} = \mathbf{I}$ and we recall that $\mathbf{Ix} = \mathbf{x}$ we obtain,

$\mathbf{x} =$ ______________________

We obtain

$$\mathbf{x} = \mathbf{A}^{-1}\mathbf{b}$$

which is the solution of the matrix equation

$$\mathbf{Ax} = \mathbf{b}$$

6-3

The evaluation of the inverse of a matrix is a very useful and important operation in the algebra of matrices. In Chapter Five we developed the necessary relationships for the evaluation of the inverse matrix using the adjoint matrix.

In order to show how the adjoint matrix is related to the inversion operation, let us multiply matrix **A** by its adjoint matrix, *adj* **A**. For example, complete the multiplication of the second order matrix **A** and its adjoint matrix by evaluating the elements of **B** in the equation **A**. *adj* **A** = **B**.

$$\begin{bmatrix} a_{11} & a_{12} \\ a_{21} & a_{22} \end{bmatrix} \begin{bmatrix} \alpha_{11} & \alpha_{21} \\ \alpha_{12} & \alpha_{22} \end{bmatrix} = \begin{bmatrix} b_{11} & b_{11} \\ b_{21} & b_{22} \end{bmatrix}$$

$b_{11} = a_{11}\alpha_{11} + a_{12}\alpha_{12}$

$b_{12} =$ ______________________________

$b_{21} =$ ______________________________

$b_{22} =$ ______________________________

$b_{11} = a_{11}\alpha_{11} + a_{12}\alpha_{12}$

$b_{12} = a_{11}\alpha_{21} + a_{12}\alpha_{22}$

$b_{21} = a_{21}\alpha_{11} + a_{22}\alpha_{12}$

$b_{22} = a_{21}\alpha_{21} + a_{22}\alpha_{22}$

We observe that the equation for the first element, b_{11}, consists of the elements of the first row of **A** multiplied by the corresponding cofactors of the same elements. We recall that b_{11} is identical to the expansion of the determinant of **A** by cofactors along the first row (see page 74), and therefore

$$b_{11} = \det \mathbf{A}$$

The element b_{12} also contains the elements of the first row but they are not multiplied by the corresponding cofactors. For the square matrix **A** of order 2 we have

$$\mathbf{A} = \begin{bmatrix} a_{11} & a_{12} \\ a_{21} & a_{22} \end{bmatrix}$$

we can calculate the specific adjoint matrix, obtaining

$$adj\ \mathbf{A} = \begin{bmatrix} a_{22} & -a_{12} \\ & \end{bmatrix}$$

complete the adjoint before continuing.

$$\text{adj}\ \mathbf{A} = \begin{bmatrix} a_{22} & -a_{12} \\ -a_{21} & a_{11} \end{bmatrix}$$

Completing the matrix multiplication we write

$$\mathbf{A} \cdot adj\ \mathbf{A} = \begin{bmatrix} a_{11} & a_{12} \\ a_{21} & a_{22} \end{bmatrix} \begin{bmatrix} a_{22} & -a_{12} \\ -a_{21} & a_{11} \end{bmatrix} = \begin{bmatrix} b_{11} & b_{12} \\ b_{21} & b_{22} \end{bmatrix}$$

Again, we note that

$$b_{11} = a_{11}a_{22} - a_{12}a_{21} = |\mathbf{A}|$$

Also we find that

$$b_{12} = a_{11}(-a_{12}) + a_{12}(a_{11}) = 0$$

$$b_{21} = a_{21}(a_{22}) + a_{22}(-a_{21}) = 0$$

$$b_{22} = a_{21}(-a_{12}) + a_{22}(a_{11}) = |\mathbf{A}|$$

In general, one may show that if the elements of a row of a matrix are multiplied by cofactors of a different row and the products are summed, the result is zero. Also, the elements on the major diagonal of the product of **A** and *adj* **A** are equal to $|\mathbf{A}|$. Therefore, for an *n*th order matrix **A**.

$$\text{(6-4)} \qquad \mathbf{A} \cdot adj\ \mathbf{A} = \begin{bmatrix} |\mathbf{A}| & 0 & \dots & 0 \\ 0 & |\mathbf{A}| & \dots & 0 \\ \cdot & \cdot & & \\ \cdot & \cdot & & \\ \cdot & \cdot & & \\ 0 & 0 & \dots & |\mathbf{A}| \end{bmatrix} = |\mathbf{A}|\ \mathbf{I}$$

Find the product **A** · adj **A** for the matrix

$$\mathbf{A} = \begin{bmatrix} 6 & 0 & 1 \\ 3 & 2 & 1 \\ 0 & 4 & 2 \end{bmatrix};$$

$$\mathbf{A}\cdot adj\,\mathbf{A} = \begin{bmatrix} |\mathbf{A}| & 0 & 0 \\ 0 & |\mathbf{A}| & 0 \\ 0 & 0 & |\mathbf{A}| \end{bmatrix} = \begin{bmatrix} 12 & 0 & 0 \\ 0 & 12 & 0 \\ 0 & 0 & 12 \end{bmatrix}$$

since

$$\det \mathbf{A} = 6 \begin{vmatrix} 2 & 1 \\ 4 & 2 \end{vmatrix} + 1 \begin{vmatrix} 3 & 2 \\ 0 & 4 \end{vmatrix} = 12$$

Rewriting equation (6-4)

$$\mathbf{A} \cdot adj\,\mathbf{A} = |\mathbf{A}| \quad \mathbf{I}$$

Dividing both sides of this equation by the determinant of **A** we obtain

(6-5) $$\mathbf{A} \frac{adj\,\mathbf{A}}{|\mathbf{A}|} = \mathbf{I}$$

Recall that equation (6-1) states

$$\mathbf{A}\mathbf{A}^{-1} = \mathbf{I}$$

and we find that the inverse of **A** may be written as

(6-6) $$\mathbf{A}^{-1} = \frac{adj\,\mathbf{A}}{|\mathbf{A}|}$$

Thus, the inverse of a square matrix is equal to the adjoint matrix divided by the determinant of the matrix.

6-5

Find the inverse of the matrix

$$\mathbf{A} = \begin{bmatrix} 4 & 1 \\ 2 & 1 \end{bmatrix}$$

$$\mathbf{A}^{-1} = \frac{adj\ \mathbf{A}}{|\mathbf{A}|} = \frac{1}{2} \begin{bmatrix} 1 & -1 \\ -2 & 4 \end{bmatrix} = \begin{bmatrix} 1/2 & -1/2 \\ -1 & 2 \end{bmatrix}$$

since

$$|\mathbf{A}| = 2$$

and

$$adj\ \mathbf{A} = \begin{bmatrix} 1 & -1 \\ -2 & 4 \end{bmatrix}$$

$$\text{Since } \mathbf{A}^{-1} = \frac{adj\ \mathbf{A}}{|\mathbf{A}|}$$

6-6

we find that the inverse of a matrix exists only if the determinant of the matrix is not equal to zero and the determinant is nonsingular. If the determinant is singular, det **A** = 0, then no inverse can be found.

In order to check the inverse determined in the previous example, show that

$$\mathbf{AA}^{-1} = \mathbf{I}$$

$$\begin{bmatrix} 4 & 1 \\ 2 & 1 \end{bmatrix} \begin{bmatrix} \quad & \quad \\ \quad & \quad \end{bmatrix} = \begin{bmatrix} \quad & \quad \\ \quad & \quad \end{bmatrix}$$

$$\begin{bmatrix} 4 & 1 \\ 2 & 1 \end{bmatrix} \begin{bmatrix} 1/2 & -1/2 \\ -1 & 2 \end{bmatrix} = \begin{bmatrix} 1 & 0 \\ 0 & 1 \end{bmatrix}$$

6-7

Find the inverse of **B** when

$$\mathbf{B} = \begin{bmatrix} 2 & 3 & 1 \\ 1 & 2 & 3 \\ 3 & 1 & 2 \end{bmatrix}$$

$$\mathbf{B}^{-1} = \begin{bmatrix} & & \\ & & \\ & & \end{bmatrix}$$

$$B^{-1} = \frac{1}{18}\begin{bmatrix} 1 & -5 & 7 \\ 7 & 1 & 5 \\ -5 & 7 & 1 \end{bmatrix}$$

If your answer was correct, proceed to Frame 6-8, otherwise follow these calculations

$$\det \mathbf{B} = \begin{vmatrix} 2 & 3 & 1 \\ 1 & 2 & 3 \\ 3 & 1 & 2 \end{vmatrix} = 2\begin{vmatrix} 2 & 3 \\ 1 & 2 \end{vmatrix} - 3\begin{vmatrix} 1 & 3 \\ 3 & 2 \end{vmatrix} + 1\begin{vmatrix} 1 & 2 \\ 3 & 1 \end{vmatrix}$$

$$= 2(1) - 3(-7) + 1(-5) = 18$$

$$\text{adj } \mathbf{B} = \begin{bmatrix} 1 & -5 & 7 \\ 7 & 1 & -5 \\ -5 & 7 & 1 \end{bmatrix} \quad \text{and } \mathbf{B}^{C} = \begin{bmatrix} 1 & 7 & -5 \\ -5 & 1 & 7 \\ 7 & -5 & 1 \end{bmatrix}$$

where, for example,

$$\alpha_{11} = (-1)^2 \begin{vmatrix} 2 & 3 \\ 1 & 2 \end{vmatrix} = 1$$

and

$$\alpha_{12} = (-1)^3 \begin{vmatrix} 1 & 3 \\ 3 & 2 \end{vmatrix} = -1(-7) = 7$$

One may check the evaluation by showing that

$$\mathbf{BB}^{-1} = \mathbf{I}$$

6-8

The method of inverting a matrix by using the adjoint matrix is not very practical for matrices of an order higher than 4 because of the large number of calculations that are necessary. For higher order matrices we utilize a digital computer to carry out all the calculations.

The properties of the inverse of a matrix are important to record. First, we note that the inverse of a nonsingular square matrix of order n is unique. Furthermore, if **A** is nonsingular, then

$$\mathbf{AB} = \mathbf{AC}$$

implies **B** = **C**. This result is obtained when we premultiply (that is, multiply on the left side of the matrices) the equation by $\mathbf{A}^{-1}$.

An obvious property of a matrix inverse is that

$$(\mathbf{A}^{-1})^{-1} = \mathbf{A}$$

when **A** is nonsingular.

Given that **I** is the identity matrix, we find that the inverse of the identity matrix is

$$\mathbf{I}^{-1} = \_\_\_\_\_\_\_\_\_\_\_\_$$

$$\underline{\mathbf{I}^{-1} = \mathbf{I}}$$

You may prove this by using the definition of the inverse, which is

$$\mathbf{A}^{-1} = \frac{adj\ \mathbf{A}}{|\mathbf{A}|}$$

Since

$$\mathbf{A} = \mathbf{I} = \begin{bmatrix} 1 & 0 & \dots & 0 \\ 0 & 1 & \dots & 0 \\ \cdot & \cdot & & \cdot \\ \cdot & \cdot & & \cdot \\ \cdot & \cdot & & \cdot \\ 0 & 0 & \dots & 1 \end{bmatrix}$$

then $|\mathbf{A}| = |\mathbf{I}| = 1$ and $adj\ \mathbf{I} = \mathbf{I}$

Therefore, we find that

$$\mathbf{I}^{-1} = \mathbf{I}$$

6-9

The inverse of a diagonal matrix

$$\mathbf{D} = \begin{bmatrix} d_{11} & 0 & \dots & 0 \\ 0 & d_{22} & \dots & 0 \\ \cdot & \cdot & & \cdot \\ \cdot & \cdot & & \cdot \\ \cdot & \cdot & & \cdot \\ 0 & 0 & \dots & d_{nn} \end{bmatrix}$$

is

$$\mathbf{D}^{-1} = \begin{bmatrix} 1/d_{11} & 0 & & 0 \\ 0 & 1/d_{22} & \dots & \cdot \\ \cdot & \cdot & & \cdot \\ \cdot & \cdot & & \cdot \\ 0 & 0 & \dots & 1/d_{nn} \end{bmatrix}$$

Verify this relationship by completing the multiplication to show $\mathbf{DD}^{-1} = \mathbf{I}$.

Clearly,

$$\mathbf{DD}^{-1} = \begin{bmatrix} 1/d_{11} & 0 & \dots & 0 \\ 0 & 1/d_{22} & \dots & 0 \\ \cdot & \cdot & & \cdot \\ \cdot & \cdot & & \cdot \\ \cdot & \cdot & & \cdot \\ 0 & 0 & \dots & 1/d_{nn} \end{bmatrix} \begin{bmatrix} d_{11} & 0 & \dots & 0 \\ 0 & d_{22} & \dots & 0 \\ \cdot & \cdot & & \cdot \\ \cdot & \cdot & & \cdot \\ \cdot & \cdot & & \cdot \\ 0 & 0 & \dots & d_{nn} \end{bmatrix} = \begin{bmatrix} 1 & 0 & \dots & 0 \\ 0 & 1 & \dots & 0 \\ \cdot & \cdot & & \cdot \\ \cdot & \cdot & & \cdot \\ \cdot & \cdot & & \cdot \\ 0 & 0 & \dots & 1 \end{bmatrix}$$

or

$$\mathbf{DD}^{-1} = \mathbf{I}$$

6-10

Consider the scalar matrix $k\mathbf{A} = \mathbf{B}$ discussed in Chapter 2. The inverse of the scalar matrix is equal to the inverse of the scalar times the inverse of the matrix. Therefore,

$$\mathbf{B}^{-1} = (k\mathbf{A})^{-1} = \frac{1}{k}\,\mathbf{A}^{-1}$$

This relationship is useful when a common factor can be removed from all the elements of a matrix **B**. Therefore, for example, the inverse **B** can be determined as follows:

$$\mathbf{B} = \begin{bmatrix} 3 & 6 \\ 3 & 9 \end{bmatrix} = 3\begin{bmatrix} 1 & 2 \\ 1 & 3 \end{bmatrix} = k\mathbf{A}$$

and

$$\mathbf{B}^{-1} = \frac{1}{3}\begin{bmatrix} 3 & -2 \\ -1 & 1 \end{bmatrix} \qquad \text{where } \det \mathbf{A} = 1$$

Find the inverse of

$$\mathbf{C} = \begin{bmatrix} 38 & 0 & 19 \\ 0 & 38 & 19 \\ 19 & 57 & 0 \end{bmatrix} \quad ; \quad \mathbf{C}^{-1} =$$

$$C^{-1} = \frac{1}{19(-8)} \begin{bmatrix} -3 & 3 & -2 \\ 1 & -1 & -2 \\ -2 & -6 & 4 \end{bmatrix}$$

since

$$C = 19 \begin{bmatrix} 2 & 0 & 1 \\ 0 & 2 & 1 \\ 1 & 3 & 0 \end{bmatrix} = 19A$$

and

$$\det A = 2(-3) + 1(-2) = -8$$

6-11

The inverse of the product of two square matrices is the product of the inverse of each matrix, but appearing in the reverse order. Therefore,

$$(AB)^{-1} = B^{-1}A^{-1} \tag{6-7}$$

Prove this relationship by utilizing the definition of the inverse operation, equation (6-1), which is

$$C^{-1}C = I$$

Let $\mathbf{C} = \mathbf{AB}$. Then

(6-8) $$(\mathbf{AB})^{-1}(\mathbf{AB}) = \mathbf{I}$$

But equation (6-7) states that

$$(\mathbf{AB})^{-1} = \mathbf{B}^{-1}\mathbf{A}^{-1}$$

Substituting equation (6-7) into (6-8) we obtain

$$\mathbf{B}^{-1}\mathbf{A}^{-1}\mathbf{AB} = \mathbf{B}^{-1}\mathbf{IB} = \mathbf{B}^{-1}\mathbf{B} = \mathbf{I}$$

which yields the desired identity.

6-12

For review purposes fill in the blanks in the following questions.

1. The inverse of a matrix $\mathbf{R}$ is written as ____________________.
2. The definition of an inverse matrix is such that $\mathbf{RR}^{-1}$ = ____________.
3. The solution to the matrix equation $\mathbf{Ry} = \mathbf{z}$ is $\mathbf{y}$ = __________________.
4. The inverse of a matrix $\mathbf{B}$ written in terms of its adjoint matrix is ________________.

1. The inverse of a matrix $\mathbf{R}$ is written as $\mathbf{R}^{-1}$.
2. The definition of an inverse matrix is such that $\mathbf{RR}^{-1} = \mathbf{I}$.
3. The solution to the matrix equation $\mathbf{Ry} = \mathbf{z}$ is $\mathbf{y} = \mathbf{R}^{-1}\mathbf{z}$.
4. The inverse of a matrix $\mathbf{B}$ written in terms of its adjoint matrix is $\mathbf{B}^{-1} = \dfrac{\text{adj }\mathbf{B}}{|\mathbf{B}|}$.

This may be a convenient stopping place. Continue when you are ready.

THE ORTHOGONAL MATRIX

6-13

A matrix defined in terms of its inverse is an orthogonal matrix. An *orthogonal matrix* is a square matrix for which its inverse is equal to its transpose, so that if

$$\mathbf{A}^{-1} = \mathbf{A}'$$

then **A** is an orthogonal matrix. Of course, one orthogonal matrix is the identity matrix.

Given a matrix

$$\mathbf{A} = \begin{bmatrix} a & -b \\ b & a \end{bmatrix}$$

find the conditions a and b must satisfy for **A** to be an orthogonal matrix.

$=1$

Since we require that

$$\mathbf{A}^{-1} = \mathbf{A}'$$

for **A** to be an orthogonal matrix, we rewrite the relationship as

$$\mathbf{I} = \mathbf{A}\mathbf{A}'$$

by premultiplying by **A**. Therefore, it is required that

$$\begin{bmatrix} a & -b \\ b & a \end{bmatrix} \begin{bmatrix} a & b \\ -b & a \end{bmatrix} = \begin{bmatrix} 1 & 0 \\ 0 & 1 \end{bmatrix}$$

Completing the multiplication, we obtain the requirement that

$$a^2 + b^2 = 1$$

for the second order matrix of the form

$$\mathbf{A} = \begin{bmatrix} a & -b \\ b & a \end{bmatrix}.$$

We note that the multiplication of the first row and second column properly yields

$$ab - ba = 0$$

6-14

Now that we have developed a method of evaluating the inverse of a matrix we may utilize the inversion process to solve a set of n simultaneous equations. The matrix representing a set of n simultaneous eauations in n unknowns is written as

$$\mathbf{A}\mathbf{x} = \mathbf{b}$$

In order to determine the unique solution we premultiply by $\mathbf{A}^{-1}$ obtaining

$$\mathbf{A}^{-1}\mathbf{A}\mathbf{x} = \mathbf{A}^{-1}\mathbf{b}$$

or

$$\mathbf{x} = \mathbf{A}^{-1}\mathbf{b}$$

Let us find the solution for the set of equations

$$3x_1 + 2x_2 = 1$$

$$2x_1 + 4x_2 = 2$$

In matrix form, the equations are written as

$$\begin{bmatrix} 3 & 2 \\ 2 & 4 \end{bmatrix} \begin{bmatrix} x_1 \\ x_2 \end{bmatrix} = \begin{bmatrix} 1 \\ 2 \end{bmatrix}$$

Since

$$\mathbf{A}^{-1} = \frac{1}{8} \begin{bmatrix} 4 & -2 \\ -2 & 3 \end{bmatrix}$$

we obtain

$$\mathbf{x} = \mathbf{A}^{-1}\mathbf{b}$$

$x_1 =$ ____________

$x_2 =$ ____________

$$x_1 = 0$$

$$x_2 = 1/2$$

since

$$\mathbf{x} = \frac{1}{8} \begin{bmatrix} 4 & -2 \\ -2 & 3 \end{bmatrix} \begin{bmatrix} 1 \\ 2 \end{bmatrix} = \frac{1}{8} \begin{bmatrix} 0 \\ 4 \end{bmatrix}$$

6-15

This method of obtaining the solution of a simultaneous set of equations is particularly useful when the unknowns $x_1, x_2, ..., x_n$ are to be evaluated for several sets of constants $b_1, b_2, ..., b_n$

Determine the solution to the example on Frame 6-14 when

$$\mathbf{b} = \begin{bmatrix} 4 \\ 1 \end{bmatrix}$$

$$\mathbf{x} = \begin{bmatrix} (14/8) \\ (-5/8) \end{bmatrix}$$

since

$$\mathbf{x} = \mathbf{A}^{-1}\mathbf{b} = \frac{1}{8}\begin{bmatrix} 4 & -2 \\ -2 & 3 \end{bmatrix}\begin{bmatrix} 4 \\ 1 \end{bmatrix} = \frac{1}{8}\begin{bmatrix} 14 \\ -5 \end{bmatrix}$$

6-16

A set of simultaneous equations is *homogeneous* if all the constant terms on the right side of the equations are equal to zero, so that

$$a_{11}x_1 + a_{12}x_2 + ... + a_{1n}x_n = 0$$

$$a_{21}x_1 + a_{22}x_2 + ... + a_{2n}x_n = 0$$

$$\vdots$$

$$a_{n1}x_1 + a_{n2}x_2 + ... + a_{nn}x_n = 0$$

Written in matrix form,

$$\mathbf{Ax} = \mathbf{b}$$

where $\mathbf{b} = \mathbf{0}$, $\mathbf{0}$ is the null or zero vector and **A** is a square matrix of order n.

It can be seen that one solution of the matrix equation

$$\mathbf{Ax} = \mathbf{0}$$

is $\mathbf{x} = \mathbf{0}$, which is called the *trivial* solution and is obtained when **A** is nonsingular.

When the coefficient matrix **A** has rank n, the trivial solution, $\mathbf{x} = \mathbf{0}$, is ______________ and the matrix **A** is ______________ .
unique/nonunique singular/nonsingular

The trivial solution, $\mathbf{x} = \mathbf{0}$, is unique and the matrix **A** is nonsingular. Since the rank is n, $n - r = 0$ and the solution is unique.

6-17

If the rank r of the matrix **A** is less than n, we can solve for r of the variables in terms of the remaining $n - r$ variables. Then there are infinitely many nontrivial solutions in this case.

A necessary and sufficient condition that a system of m homogeneous equations in n unknowns

$$\mathbf{Ax} = \mathbf{0}$$

have nontrival solutions is that the rank, r, of **A** be less than n. When $m = n$ and the **A** matrix is square, a nontrivial solution is obtained if and only if **A** is singular.

A set of equations

$$4x_1 + 2x_2 = 0$$

$$3x_1 + 3x_2 = 0$$

is represented by the matrix equation

$$\begin{bmatrix} 4 & 2 \\ 3 & 3 \end{bmatrix} \mathbf{x} = \begin{bmatrix} 0 \\ 0 \end{bmatrix}$$

The solution to these equations is $x_1 =$ ______________ and $x_2 =$ ______________ and the solution is a ______________ solution.
trivial/nontrivial

The solution to the equations is $x_1 = 0$ and $x_2 = 0$ and the solution is a trivial solution.

This is the case since det $\mathbf{A} = 6$; thus $\mathbf{A}$ is a nonsingular square matrix of rank 2.

6-18

A set of equations

$$4x_1 + 2x_2 = 0$$

$$2x_1 + x_2 = 0$$

is represented by the matrix equation (fill in the matrices):

$$\begin{bmatrix} \quad & \quad \\ \quad & \quad \end{bmatrix} \mathbf{x} = \begin{bmatrix} \quad \\ \quad \end{bmatrix}$$

The solution to these equations is $x_1 =$ ______________ and $x_2 =$ ______________ and the solution is a ______________ solution.
trivial/nontrivial

The solution to these equations is $x_1 = -\beta/2$ and $x_2 = \beta$ (where β = arbitrary constant) and the solution is a nontrivial solution.

$$\begin{bmatrix} 4 & 2 \\ 2 & 1 \end{bmatrix} \mathbf{x} = \mathbf{0}$$

and therefore det $\mathbf{A} = 0$. Solving the second equation for x_1 we have

$$2x_1 = x_2$$

or

$$x_1 = \frac{-x_2}{2}$$

Then, setting x_2 equal to the arbitrary constant β, we have

$$x_1 = -\beta/2$$

or

$$\mathbf{x} = \beta \begin{bmatrix} -1/2 \\ 1 \end{bmatrix}$$

Since β is arbitrary an infinite number of solutions exists. Recall from Chapter 5 that since the rank of $\mathbf{A}$ is 1 we will have one aribtrarily established unknown and one unknown uniquely determined.

SUMMARY

1. The inverse of a square matrix **A** is written as $\mathbf{A}^{-1}$ and it satisfies the relationship

$$\mathbf{A}^{-1}\mathbf{A} = \mathbf{A}\mathbf{A}^{-1} = \mathbf{I}$$

where the product of the matrix and its inverse are commutative.

2. The matrix equation representing a set of n simultaneous equations in n unknowns is written as

$$\mathbf{A}\mathbf{x} = \mathbf{b}$$

and the unique solution

$$\mathbf{x} = \mathbf{A}^{-1}\mathbf{b}$$

is obtained when the matrix **A** is nonsingular.

3. The inverse of a matrix **A** is

$$\mathbf{A}^{-1} = \frac{adj\ \mathbf{A}}{|\mathbf{A}|}$$

where *adj* **A** is the adjoint of **A** and $|\mathbf{A}|$ is the determinant of **A**. The inverse exists only when the determinant of **A** is not equal to zero.

4. If **A** is nonsingular, then

$$\mathbf{AB} = \mathbf{AC}$$

implies **B** = **C**.

5. The inverse of a diagonal matrix

$$\mathbf{D} = \begin{bmatrix} d_{11} & 0 & \cdots & 0 \\ 0 & d_{22} & & 0 \\ \vdots & \vdots & & \vdots \\ 0 & 0 & \cdots & d_{nn} \end{bmatrix}$$

is

$$\begin{bmatrix} 1/d_{11} & 0 & \ldots & 0 \\ 0 & 1/d_{22} & \ldots & 0 \\ \vdots & \vdots & & \vdots \\ 0 & 0 & \ldots & 1/d_{nn} \end{bmatrix}$$

6. The inverse of the scalar matrix equals the inverse of the scalar times the inverse of the matrix. Therefore,

$$(k\mathbf{A})^{-1} = \frac{1}{k}\,\mathbf{A}^{-1}$$

7. The inverse of the product of two square matrices is the product of the inverse of each matrix appearing in reverse order. Therefore,

$$(\mathbf{AB})^{-1} = \mathbf{B}^{-1}\mathbf{A}^{-1}$$

8. An orthogonal matrix is a square matrix for which its inverse is equal to its transpose so that if

$$\mathbf{A}^{-1} = \mathbf{A}'$$

then **A** is an orthogonal matrix. This equation may also be written as

$$\mathbf{AA}' = \mathbf{I}$$

9. A set of simultaneous equations is homogeneous if all the constant terms on the right side of the equations equal zero, so that the matrix equation representing these equations is

$$\mathbf{Ax} = \mathbf{0}$$

When the matrix **A** is nonsingular the solution to this equation is

$$\mathbf{x} = \mathbf{0}$$

and is called the trivial solution.

When the matrix **A** is singular a nontrivial solution is obtained and there is an infinite number of solutions.

EXERCISES

1. Find the inverse of the product of the two matrices **D** and **P**.

$$(\mathbf{DP})^{-1} = \_\_\_\_\_\_\_\_\_\_\_\_\_\_\_\_$$

2. Find the inverse of the matrix **Q** where

$$\mathbf{Q} = \begin{bmatrix} 0 & 23 & 69 \\ 46 & 0 & 23 \\ 0 & 46 & 69 \end{bmatrix} \qquad \mathbf{Q}^{-1} = \begin{bmatrix} & & \\ & & \\ & & \end{bmatrix}$$

3. Find the inverse of

$$\mathbf{P} = \begin{bmatrix} 18 & 0 & 0 & 0 \\ 0 & 21 & 0 & 0 \\ 0 & 0 & 13 & 0 \\ 0 & 0 & 0 & 5 \end{bmatrix} \qquad \mathbf{P}^{-1} =$$

4. The definition of the inverse of the matrix **D** is such that

$$\_\_\_\_\_\_\_\_\_\_\_\_ = \mathbf{I}$$

5. Find the inverse of the matrix **A** when

$$\mathbf{A} = \begin{bmatrix} 8 & 4 & 2 \\ 2 & 8 & 4 \\ 1 & 2 & 8 \end{bmatrix} \qquad \mathbf{A}^{-1} = \begin{bmatrix} & & \\ & & \\ & & \end{bmatrix}$$

6. Solve the set of simultaneous equations

$$8x_1 + 4x_2 + 2x_3 = 98$$
$$2x_1 + 8x_2 + 4x_3 = 196$$
$$x_1 \;\; 3\; 2x_2 + 8x_3 = 14$$

using the inverse of **A** obtained in problem 5.

$$\mathbf{x} = \begin{bmatrix} \quad \\ \quad \\ \quad \end{bmatrix}$$

7. If **P** is a nonsingular matrix and

$$\mathbf{PA} = \mathbf{BP}$$

then **A** = **B** is ________.
true/false

8. Determine the solution of the matrix equation

$$\mathbf{Ax} = \mathbf{0}$$

where

$$\mathbf{A} = \begin{bmatrix} 1 & 0 & 1 \\ 0 & 1 & -2 \\ 1 & 0 & -1 \end{bmatrix}$$

9. Determine the solution of the matrix equation

$$\mathbf{Ax} = \mathbf{0}$$

where

$$\mathbf{A} = \begin{bmatrix} 1 & 0 & 1 \\ 0 & 1 & -1 \\ 1 & 0 & 1 \end{bmatrix}$$

10. Prove that **Q** is an orthogonal matrix when

$$\mathbf{Q} = \begin{bmatrix} \frac{1}{\sqrt{3}} & \frac{1}{\sqrt{6}} & \frac{-1}{\sqrt{2}} \\ \frac{1}{\sqrt{3}} & \frac{-2}{\sqrt{6}} & 0 \\ \frac{1}{\sqrt{3}} & \frac{1}{\sqrt{6}} & \frac{1}{\sqrt{2}} \end{bmatrix}$$

The answers appear on page 243.

7 the characteristic equation of a matrix

7-1

In many important applications of matrices in the physical science, engineering, and the social sciences, the characteristic equation and its associated roots must be considered.

Given a square matrix **A** of order n, the problem is to determine scalars λ and the corresponding nonzero vectors **x** that simultaneously satisfy the equation.

(7-1) $$\mathbf{Ax} = \lambda \mathbf{x}$$

This is known as the *characteristic value problem,* often called the *eigenvalue problem.* The question can be stated: is there a scalar number λ, which when multiplying the vector **x**, yields the identical vector resulting from the product **Ax**?

Adding $\lambda\mathbf{x}$ to both sides of equation (7-1) we obtain

(7-2) $$\mathbf{Ax} - \lambda \mathbf{x} = \mathbf{0}$$

Then, rewriting equation (7-2) by removing **x** on the right side we obtain the homogeneous equation

(7-3) $$(\mathbf{A} - \lambda \mathbf{I})\, \mathbf{x} = \mathbf{0}$$

where **I** = identity matrix.

As we found in Chapter 6, the homogeneous equation (7-3) has a nontrivial solution if and only if det (____________) = ________

$$(\mathbf{A} - \lambda \mathbf{I})\, \mathbf{x} = \mathbf{0}$$

has a nontrivial solution if and only if

$$\det (\mathbf{A} - \lambda \mathbf{I}) = 0$$

See page 140 for review if your answer was incorrect.

The polynomial equation in λ resulting from

$$\det(\mathbf{A} - \lambda\mathbf{I})$$

is called the *characteristic polynomial.*

Given a matrix

$$\mathbf{A} = \begin{bmatrix} 3 & 2 \\ 4 & 1 \end{bmatrix}$$

let us obtain its characteristic polynomial,

$$\det(\mathbf{A} - \lambda\mathbf{I}) = \det\left(\begin{bmatrix} 3 & 2 \\ 4 & 1 \end{bmatrix} - \lambda\begin{bmatrix} 1 & 0 \\ 0 & 1 \end{bmatrix}\right)$$

$$= \begin{bmatrix} (3-\lambda) & 2 \\ 4 & (1-\lambda) \end{bmatrix} = (3-\lambda)(1-\lambda)-8$$

$$= \lambda^2 - 4\lambda - 5$$

Determine the characteristic polynomial for the matrix

$$\mathbf{A} = \begin{bmatrix} 1 & 2 \\ 2 & 1 \end{bmatrix} \qquad \det(\mathbf{A} - \lambda\mathbf{I}) = \_\_\_\_\_\_\_\_\_\_$$

$$\det(\mathbf{A} - \lambda\mathbf{I}) = \underline{\lambda^2 - 2\lambda - 3}$$

since

$$\det(\mathbf{A} - \lambda\mathbf{I}) = \begin{vmatrix} (1-\lambda) & 2 \\ 2 & (1-\lambda) \end{vmatrix} = (1-\lambda)^2 - 4 = \lambda^2 - 2\lambda - 3$$

7-3

In order to find the roots of the characteristic polynomial, we set it equal to zero, obtaining

$$\det(\mathbf{A} - \lambda\,\mathbf{I}) = 0$$

which is called the *characteristic equation.* We note that for a matrix **A** of order *n* we obtain a characteristic equation of order *n* which contains *n* roots. The characteristic equation obtained for

$$\mathbf{A} = \begin{bmatrix} 1 & 2 \\ 2 & 1 \end{bmatrix}$$

is

$$\det(\mathbf{A} - \lambda\mathbf{I}) = 0$$

or

$$\lambda^2 - 2\lambda - 3 = 0$$

The roots, or factors, of the characteristic equation are called the *characteristic roots,* and are sometimes called the *eigenvalues.* In this case the characteristic roots are

$$\lambda_1 = -1, \; \lambda_2 = 3$$

since

$$(\lambda+1)\;(\lambda-3) = \lambda^2 - 2\lambda - 3 = 0$$

Consider the matrix

$$A = \begin{bmatrix} 4 & -5 \\ 2 & -3 \end{bmatrix}$$

Obtain the characteristic equation and the characteristic roots.

$$\det(A-\lambda I) = \begin{vmatrix} (4-\lambda) & -5 \\ 2 & (-3-\lambda) \end{vmatrix} = \lambda^2 - \lambda - 2 = 0$$

is the characteristic equation. The roots are

$$\lambda_1 = 2 \text{ and } \lambda_2 = -1$$

since

$$(\lambda-2)(\lambda+1) = \lambda^2 - \lambda - 2 = 0$$

7-4

Lets not lose sight of the original problem, which is to determine scalars λ associated with a square matrix **A** of order n that satisfy the equation

__________________ .

This problem is known as the ________________________ problem.

The problem is to determine scalars λ associated with square matrix **A** of order n that satisfy the equation $\mathbf{Ax} = \lambda\mathbf{x}$.

This problem is known as the *characteristic value* problem (or *eigenvalue* problem).

7-5

The equation det (   ) = 0 is called the ________________ equation and it possesses __________ roots for a matrix of order n.

The equation det $(\mathbf{A}-\lambda\mathbf{I}) = 0$ is called the *characteristic* equation and possesses n roots for a matrix of order n.

7-6

Find the characteristic equation and characteristic roots for the matrix

$$\mathbf{A} = \begin{bmatrix} 1 & 1 & -2 \\ -1 & 2 & 1 \\ 0 & 1 & -1 \end{bmatrix}$$

det $(\mathbf{A} - \lambda\mathbf{I}) = 0 =$ ________________________

$\lambda_1 =$ ________ , $\lambda_2 =$ ________ , $\lambda_3 =$ ________

$$\det(\mathbf{A}-\lambda\mathbf{I}) = 0 = -\lambda^3 + 2\lambda^2 + \lambda - 2$$

$$\lambda_1 = -1 \quad , \quad \lambda_2 = 2 \quad , \quad \lambda_3 = 1$$

Read the following if you had an error; otherwise, proceed to Frame 7-7.

$$\det(\mathbf{A}-\lambda\mathbf{I}) = \begin{vmatrix} (1-\lambda) & 1 & -2 \\ -1 & (2-\lambda) & 1 \\ 0 & 1 & (-1-\lambda) \end{vmatrix}$$

$$= (1-\lambda)\ \big((2-\lambda)(-1-\lambda)-1\big)\ + (-1)(-1)^3\ \big((-1-\lambda)+2\big)$$

$$= (1-\lambda)(\lambda^2-\lambda-3) + (-\lambda+1)$$

$$= -\lambda^3 + 2\lambda^2 + \lambda - 2 = 0$$

Factoring $\det(\mathbf{A}-\lambda\mathbf{I}) = 0$, we have

$$\lambda_1 = -1,\ \lambda_2 = 2,\ \lambda_3 = 1$$

and

$$-(\lambda+1)(\lambda-2)(\lambda-1) = \lambda^3 + 2\lambda^2 + \lambda - 2 = 0$$

7-7

In general it can be shown that for a square matrix of order n

$$\det(\mathbf{A}-\lambda\mathbf{I}) = \lambda^n + c_{n-1}\ \lambda^{n-1} + c_{n-2}\ \lambda^{n-2} + \ldots + c_1\lambda + (-1)^n\ |\mathbf{A}|$$

where we obtain an nth order polynomial and c_k is the kth coefficient of the polynomial. The last term is a constant equal to $(-1)^n$ times the value of the determinant of **A**.

If the roots of the characteristic equation are not integers, as in the examples in the preceding pages, the task of ascertaining them is much more difficult. It is typical to utilize a numerical method such as the Newton-Raphson method. When the order of the characteristic equation is higher than 4, a digital computer program for solving the roots of a polynomial is usually employed.

Corresponding to each characteristic root, λ_i, of the characteristic equation there is a *characteristic vector* $\mathbf{x}_i$. This characteristic vector is the solution of the homogeneous equations that are obtained from

$$(\mathbf{A}-\lambda_i \mathbf{I})\ \mathbf{x}_i = \mathbf{0}$$

The boldface vector $\mathbf{x}_i$ is the vector corresponding to the *i*th root.

$$\mathbf{x}_i = \begin{bmatrix} x_1 \\ x_2 \\ \cdot \\ \cdot \\ \cdot \\ x_n \end{bmatrix}$$

Let us consider the matrix

$$\mathbf{A} = \begin{bmatrix} 1 & 2 \\ 2 & 1 \end{bmatrix}$$

with the characteristic roots $\lambda_1 = -1, \lambda_2 = 3$ discussed in Frame 7-3. Associated with the first root we have

$$(\mathbf{A}-\lambda_i\mathbf{I})\ \mathbf{x}_1 = \mathbf{0}$$

or

$$\left(\begin{bmatrix} 1 & 2 \\ 2 & 1 \end{bmatrix} - (-1) \begin{bmatrix} 1 & 0 \\ 0 & 1 \end{bmatrix} \right) \begin{bmatrix} x_1 \\ x_2 \end{bmatrix} = \begin{bmatrix} 0 \\ 0 \end{bmatrix}$$

which implies that

$$\begin{bmatrix} 2 & 2 \\ 2 & 2 \end{bmatrix} \begin{bmatrix} x_1 \\ x_2 \end{bmatrix} = \begin{bmatrix} 0 \\ 0 \end{bmatrix}$$

$\mathbf{x}_1 = \begin{bmatrix} x_1 \\ x_2 \end{bmatrix}$ is the characteristic vector associated with the root

________________________.

$x_1 = \begin{bmatrix} x_1 \\ x_2 \end{bmatrix}$ is the characteristic vector associated with the root $\lambda_1 = -1$.

7-8

The elements of $\mathbf{x}_1$ are x_1 and x_2. How are they related to each other?

$x_1 =$ ________________________

$$x_1 = -x_2 \qquad \text{since} \qquad 2x_1 + 2x_2 = 0$$

is obtained from either row of the equation $(\mathbf{A} - \lambda_1\mathbf{I})\mathbf{x}_1 = \mathbf{0}$

7-9

Let $x_1 = \beta$, an arbitrary constant. Find $\mathbf{x}_1 = \beta \begin{bmatrix} \ \\ \ \end{bmatrix}$

$$\mathbf{x}_1 = \beta \begin{bmatrix} 1 \\ -1 \end{bmatrix}$$

The subscript 1 on $\mathbf{x}_1$ implies that we have the characteristic vector $\mathbf{x}$ associated with the first root $\lambda_1 = -1$. Note that the matrix $(\mathbf{A}-\lambda_1\mathbf{I})$ has a rank of 1 and there is one arbitrary value and one unique value in $\mathbf{x}$.

7-10

The second characteristic root is $\lambda_2 = +3$. Therefore,

$$(\mathbf{A}-\lambda_2\mathbf{I})\mathbf{x}_2 = 0$$

and

$$\left(\begin{bmatrix} 1 & 2 \\ 2 & 1 \end{bmatrix} - (3)\begin{bmatrix} 1 & 0 \\ 0 & 1 \end{bmatrix}\right)\begin{bmatrix} x_1 \\ x_2 \end{bmatrix} = \begin{bmatrix} 0 \\ 0 \end{bmatrix}$$

or

$$\begin{bmatrix} -2 & 2 \\ 2 & -2 \end{bmatrix}\begin{bmatrix} x_1 \\ x_2 \end{bmatrix} = \begin{bmatrix} 0 \\ 0 \end{bmatrix}$$

Recall here that

$$\mathbf{x}_2 = \begin{bmatrix} x_1 \\ x_2 \end{bmatrix}$$

For this root λ_2, how is x_1 related to x_2?

$x_1 =$ ____________________

$x_1 = x_2$

7-11

Let $x_1 = \gamma$, an arbitrary constant in order to obtain

$$\mathbf{x}_2 = \gamma \begin{bmatrix} \\ \\ \end{bmatrix}$$

$$\mathbf{x}_2 = \gamma \begin{bmatrix} 1 \\ \\ 1 \end{bmatrix}$$

Again with λ_2

$(\mathbf{A} - \lambda_2\mathbf{I}) = \begin{bmatrix} -2 & 2 \\ 2 & -2 \end{bmatrix}$ is of rank 1 and there is one arbitrary value in $\mathbf{x}_2$ and one uniquely determined value.

7-12

For the matrix

$$\mathbf{A} = \begin{bmatrix} 1 & 1 & -2 \\ -1 & 2 & 1 \\ 0 & 1 & -1 \end{bmatrix}$$

we found (page 153) that the characteristic roots of the matrix are

$$\lambda_1 = -1, \lambda_2 = 2, \quad \lambda_3 = 1$$

Associated with the first root we have

$$\left(\begin{bmatrix} 1 & 1 & -2 \\ -1 & 2 & 1 \\ 0 & 1 & -1 \end{bmatrix} - (-1) \begin{bmatrix} 1 & 0 & 0 \\ 0 & 1 & 0 \\ 0 & 0 & 1 \end{bmatrix} \right) \begin{bmatrix} x_1 \\ x_2 \\ x_3 \end{bmatrix} = \begin{bmatrix} 0 \\ 0 \\ 0 \end{bmatrix}$$

or

$$\begin{bmatrix} 2 & 1 & -2 \\ -1 & 3 & 1 \\ 0 & 1 & 0 \end{bmatrix} \begin{bmatrix} x_1 \\ x_2 \\ x_3 \end{bmatrix} = \begin{bmatrix} 0 \\ 0 \\ 0 \end{bmatrix}$$

Note that this matrix is of rank 2 and we will set one variable arbitrarily. The last row implies that $x_2 = 0$ and the first or second row yields the equation

$$x_3 = \underline{\hspace{6cm}}$$

$$x_3 = x_1$$

since

$$2x_1 - 2x_3 = 0$$

Then, selecting $x_1 = \beta$, an arbitrary constant, we obtain

$$\mathbf{x}_1 = \beta \begin{bmatrix} 1 \\ 0 \\ 1 \end{bmatrix}$$

7-13

Determine the characteristic vectors associated with the second and third roots using the arbitrary constants γ and δ for the second and third vectors respectively.

$$\mathbf{x}_2 = \gamma \begin{bmatrix} 1 \\ 3 \\ 1 \end{bmatrix} \qquad \mathbf{x}_3 = \delta \begin{bmatrix} 3 \\ 2 \\ 1 \end{bmatrix}$$

If you were correct, skip to Frame 7-14; if not, continue reading here.

For the second root, $\lambda_2 = 2$ we obtain

$$\left(\begin{bmatrix} 1 & 1 & -2 \\ -1 & 2 & 1 \\ 0 & 1 & -1 \end{bmatrix} - (2) \begin{bmatrix} 1 & 0 & 0 \\ 0 & 1 & 0 \\ 0 & 0 & 1 \end{bmatrix} \right) \begin{bmatrix} x_1 \\ x_2 \\ x_3 \end{bmatrix} = \begin{bmatrix} 0 \\ 0 \\ 0 \end{bmatrix}$$

or

$$\begin{bmatrix} -1 & 1 & -2 \\ -1 & 0 & 1 \\ 0 & 1 & -3 \end{bmatrix} \begin{bmatrix} x_1 \\ x_2 \\ x_3 \end{bmatrix} = \begin{bmatrix} 0 \\ 0 \\ 0 \end{bmatrix}$$

This matrix is of rank 2. The third row yields the equation

$$x_2 = 3x_3$$

The second row yields

Setting $x_3 = \gamma$, we obtain

$$x_2 = \gamma \begin{bmatrix} 1 \\ 3 \\ 1 \end{bmatrix}$$

For the third root $\lambda_3 = 1$ we obtain

$$\begin{bmatrix} 0 & 1 & -2 \\ -1 & 1 & 1 \\ 0 & 1 & -2 \end{bmatrix} \begin{bmatrix} x_1 \\ x_2 \\ x_3 \end{bmatrix} = \begin{bmatrix} 0 \\ 0 \\ 0 \end{bmatrix}$$

Again this matrix is of rank 2. The third row yields

$$x_2 = 2x_3$$

The second row yields

$$x_1 = x_2 + x_3$$

or

$$x_1 = 3x_3$$

Setting x_3 equal to the arbitrary constant δ, we obtain

$$x_3 = \delta \begin{bmatrix} 3 \\ 2 \\ 1 \end{bmatrix}$$

7-14

The characteristic value problem is to find a scalar λ_i and its associated vector $\mathbf{x}_i$ that satisfy the equation $\mathbf{A}\mathbf{x}_i = \lambda_i \mathbf{x}_i$.

For the last problem verify that for $\lambda_3 = 1$ and

$$\mathbf{x}_3 = \begin{bmatrix} 3 \\ 2 \\ 1 \end{bmatrix}$$

$\mathbf{A}\mathbf{x}_3 = \lambda_3 \mathbf{x}_3$, where we have selected $\delta = 1$ for convenience.

$\mathbf{A}\mathbf{x}_3$ yields

$$\begin{bmatrix} 1 & 1 & -2 \\ -1 & 2 & 1 \\ 0 & 1 & -1 \end{bmatrix} \begin{bmatrix} 3 \\ 2 \\ 1 \end{bmatrix} = \begin{bmatrix} 3 \\ 2 \\ 1 \end{bmatrix}$$

which is identically equal to $\lambda_3 \mathbf{x}_3$ since $\lambda_3 = 1$.

7-15

Symmetric matrices occur quite often in the application of matrix algebra to physical and business problems. The solution of the characteristic value problem for symmetric matrices has a valuable property.

The characteristic roots of a symmetric matrix with real elements are all real. Note that the roots for

$$\mathbf{A} = \begin{bmatrix} 1 & 2 \\ 2 & 1 \end{bmatrix}$$

considered on page 150 were found to be -1 and 3.

Find the characteristic roots for the symmetric matrix **B** where

$$\mathbf{B} = \begin{bmatrix} 0.6 & 0.8 \\ 0.8 & -0.6 \end{bmatrix}$$

$\lambda_i =$ ______, $\lambda_2 =$ ______

$\lambda_1 = +1$, $\lambda_2 = -1$ is the correct answer because the characteristic equation is $\lambda^2 - 1 = 0$.

7-16

Thus far we have considered square matrices, **A**, which result in distinct characteristic roots. Let us now consider matrix **A** that possesses repeated characteristic roots. The characteristic equation of the matrix

$$\mathbf{A} = \begin{bmatrix} 2 & 2 & 1 \\ 1 & 3 & 1 \\ 1 & 2 & 2 \end{bmatrix}$$

is

$$\det(\mathbf{A} - \lambda\mathbf{I}) = \lambda^3 - 7\lambda^2 + 11\lambda - 5$$
$$= (\lambda-1)(\lambda-1)(\lambda-5) = 0$$

Therefore, the roots are

$$\lambda_1 = 1, \lambda_2 = 1, \lambda_3 = 5$$

For $\lambda_1 = 1$, we obtain

$$(\mathbf{A} - \lambda_1 \mathbf{I})\mathbf{x}_1 = \mathbf{0}$$

or

$$\begin{bmatrix} 1 & 2 & 1 \\ 1 & 2 & 1 \\ 1 & 2 & 1 \end{bmatrix} \begin{bmatrix} x_1 \\ x_2 \\ x_3 \end{bmatrix} = \begin{bmatrix} 0 \\ 0 \\ 0 \end{bmatrix}$$

The rank of this matrix is ___ (2/1) and, therefore, there ______________ (is one/are two) arbitrarily selected variable(s).

The rank of the matrix is 1 and therefore there are two arbitrarily selected variables.

7-17

Therefore, any row yields

$$x_1 + 2x_2 + x_3 = 0$$

If we select $x_3 = \beta$ and $x_2 = 0$, we obtain

$$x_1 = \beta$$

One solution is

$$\mathbf{x}_1 = \beta \begin{bmatrix} -1 \\ 0 \\ 1 \end{bmatrix}$$

For the root $\lambda_2 = 1$, we obtain the same matrix equation as obtained for λ_1, $= 1$ and

$$\begin{bmatrix} 1 & 2 & 1 \\ 1 & 2 & 1 \\ 1 & 2 & 2 \end{bmatrix} \begin{bmatrix} x_1 \\ x_2 \\ x_3 \end{bmatrix} = \begin{bmatrix} 0 \\ 0 \\ 0 \end{bmatrix}$$

Again we have $x_1 + 2x_2 + x_3 = 0$ and we may select two arbitrary variables.

Selecting $x_3 = 0$, then

$x_1 =$ ____________________

$$x_1 = -2x_2$$

Selecting $x_2 = \gamma$, we have

$$\mathbf{x}_2 = \gamma \begin{bmatrix} -2 \\ 1 \\ 0 \end{bmatrix}$$

There are an infinite number of solutions to the matrix of rank 1. You could select another set of arbitrary variables and obtain another answer.

7-18

Determine the remaining vector corresponding to $\lambda_3 = 5$.

$$\mathbf{x}_3 = \delta \begin{bmatrix} \quad \end{bmatrix}$$

$$\mathbf{x}_3 = \delta \begin{bmatrix} 1 \\ 1 \\ 1 \end{bmatrix}$$

since

$$(\mathbf{A} - \lambda_3 \mathbf{I})\mathbf{x}_3 = \begin{bmatrix} -3 & 2 & 1 \\ 1 & -2 & 1 \\ 1 & 2 & -3 \end{bmatrix} \begin{bmatrix} x_1 \\ x_2 \\ x_3 \end{bmatrix} = \begin{bmatrix} 0 \\ 0 \\ 0 \end{bmatrix}$$

which is of rank 2. Setting $x_3 = \delta$ and subtracting the second row from the first, we have

$$-4x_1 + 4x_2 = 0, \text{ or } x_1 = x_2$$

Adding the second row to the third, we obtain

$$2x_1 - 2x_3 = 0, \text{ or } x_1 = x_3$$

Therefore,

$$x_1 = x_2 = x_3 = \delta$$

This is a convenient stopping place in Chapter Seven. Proceed when you are ready.

LINEAR DEPENDENCE

7-19

Let us consider a set of vectors in which all the vectors have n elements. The vectors in a set $(\mathbf{x}_1, \mathbf{x}_2, ..., \mathbf{x}_n)$ are said to be *linearly dependent* if there exists a relationship among the vectors such that

$$k_1\mathbf{x}_1 + k_2\mathbf{x}_2 + ... + k_n\mathbf{x}_n = \mathbf{0}$$

where the k_i are scalar numbers, and at least one is not zero. If no such relationship exists, the vectors are *linearly independent.*

Consider the two vectors

$$\mathbf{x}_1 = \begin{bmatrix} 2 \\ 3 \end{bmatrix}, \quad \mathbf{x}_2 = \begin{bmatrix} -6 \\ -9 \end{bmatrix}$$

The vectors are linearly ________________ .
dependent/independent

The vectors are linearly dependent since

$$3\mathbf{x}_1 + \mathbf{x}_2 = \mathbf{0}$$

Note that one vector is proportional to the other.

7-20

The vectors $\mathbf{x}_1, \mathbf{x}_2, ..., \mathbf{x}_m$ are linearly dependent if an only if the rank of the matrix $\mathbf{M} = [\mathbf{x}_1, \mathbf{x}_2, ..., \mathbf{x}_m]$, with the given vectors as columns, is less than m. They are independent if and only if the rank m is equal to m. Therefore, for two the vectors,

$$\mathbf{x}_1 = \begin{bmatrix} 2 \\ 3 \end{bmatrix} \quad \mathbf{x}_2 = \begin{bmatrix} -6 \\ -9 \end{bmatrix}$$

we have

$$\mathbf{M} = \begin{bmatrix} 2 & -6 \\ 3 & -9 \end{bmatrix}$$

and since det $\mathbf{M} = 0$, the rank of $\mathbf{M}$ is equal to 1, while $m = 2$, so the vectors are dependent.

Consider the three characteristic vectors obtained on pages 157-160 for the matrix

$$\mathbf{A} = \begin{bmatrix} 1 & 1 & -2 \\ -1 & 2 & 1 \\ 0 & 1 & -1 \end{bmatrix}$$

$$\mathbf{x}_1 = \begin{bmatrix} 1 \\ 0 \\ 1 \end{bmatrix}, \quad \mathbf{x}_2 = \begin{bmatrix} 11 \\ 3 \\ 1 \end{bmatrix}, \quad \mathbf{x}_3 = \begin{bmatrix} 3 \\ 2 \\ 1 \end{bmatrix}$$

where $\beta = \gamma = \delta = 1$.

The set of vectors $\mathbf{M} = [\mathbf{x}_1, \mathbf{x}_2, \mathbf{x}_3]$ is linearly ________________ .

The set of vectors $\mathbf{M} = [\mathbf{x}_1, \mathbf{x}_2, \mathbf{x}_3]$, is linearly independent.

$$\mathbf{M} = \begin{bmatrix} 1 & 1 & 3 \\ 0 & 3 & 2 \\ 1 & 1 & 1 \end{bmatrix}, \quad \det \mathbf{M} = 1(1) + 1(-7) = -6,$$

and the rank of **M** is equal to 3, which equals m, the number of column vectors

7-21

Consider the set of vectors obtained on pages 163-165 for

$$\mathbf{A} = \begin{bmatrix} 2 & 2 & 1 \\ 1 & 3 & 1 \\ 1 & 2 & 2 \end{bmatrix}; \quad \mathbf{x}_1 = \begin{bmatrix} -1 \\ 0 \\ 1 \end{bmatrix}, \quad \mathbf{x}_2 = \begin{bmatrix} -2 \\ 1 \\ 0 \end{bmatrix}, \quad \mathbf{x}_3 = \begin{bmatrix} 1 \\ 1 \\ 1 \end{bmatrix}$$

The set of vectors $[\mathbf{x}_1, \mathbf{x}_2, \mathbf{x}_3]$ is linearly ________________ .

The set of vectors $[\mathbf{x}_1, \mathbf{x}_2, \mathbf{x}_3]$ is linearly independent.

$$\mathbf{M} = \begin{bmatrix} -1 & -2 & 1 \\ 0 & 1 & 1 \\ 1 & 0 & 1 \end{bmatrix}, \det \mathbf{M} = -1(1) + 1(-3) = -4.$$

The rank of **M** is equal to 3, which is equal to m.

7-22

Four vectors, $\mathbf{x}_1, \mathbf{x}_2, \mathbf{x}_3$, and $\mathbf{x}_4$ are independent if and only if the rank of the matrix $\mathbf{M} = [\mathbf{x}_1, \mathbf{x}_2, \mathbf{x}_3, \mathbf{x}_4]$ is equal to ______________________________

Four vectors are independent if and only if the rank of the matrix **M** is equal to 4.

7-23

For the matrix

$$\mathbf{C} = \begin{bmatrix} 1 & 2 \\ 3 & 2 \end{bmatrix}$$

obtain the characteristic roots and the matrix **M**. Determine if the characteristic vectors are independent.

The characteristic equation for the matrix **C** is

$$\det(\mathbf{C}-\lambda\mathbf{I}) = \lambda^2 - 3\lambda - 4 = (\lambda + 1)(\lambda - 4) = 0$$

For the first root, $\lambda_1 = -1$, we obtain

$$\mathbf{x}_1 = \beta \begin{bmatrix} -1 \\ 1 \end{bmatrix}$$

where β = arbitrary number.

For the second root, $\lambda_2 = 4$, we have

$$\mathbf{x}_2 = \gamma \begin{bmatrix} 2/3 \\ 1 \end{bmatrix}$$

where γ = arbitrary number.

The matrix **M** consists of the characteristic vectors as follows:

$$\mathbf{M} = [\mathbf{x}_1, \mathbf{x}_2] = \begin{bmatrix} -1 & 2/3 \\ 1 & 1 \end{bmatrix}$$

where we have set $\gamma = \beta = 1$ in this case.

In order to determine if the characteristic vectors are independent we evaluate the determinant of the matrix **M** as follows:

$$\det \mathbf{M} = \det \begin{bmatrix} -1 & 2/3 \\ 1 & 1 \end{bmatrix} = -\frac{5}{3}$$

Since the determinant of the matrix **M** is nonzero, the characteristic vectors are independent.

This is a convenient stopping place.

7-24

In general we may state that a square matrix **A** is singular if and only if its columns (or rows) are linearly dependent.

When the characteristic roots of a matrix **A** are distinct and the associated characteristic vectors are nonzero, then the vectors $\mathbf{x}_1, \mathbf{x}_2, \mathbf{x}_3 \ldots \mathbf{x}_n$ associated with the characteristic roots are linearly independent.

Utilizing the definition of the determinant of a transposed matrix, show that the characteristic roots of **A** and **A**′ are the same.

The characteristic roots of **A** are obtained from

$$\det(\mathbf{A} - \lambda\mathbf{I}) = 0.$$

The characteristic roots of **A**′ are obtained from

$$\det(\mathbf{A}' - \lambda\mathbf{I}) = 0.$$

Since **I** is not changed by transposition, we can write this equation as

$$|(\mathbf{A} - \lambda\mathbf{I})'| = 0$$

However, we recall from Chapter 4 that

$$|\mathbf{B}| = |\mathbf{B}'|$$

and, therefore we have shown that

$$|\mathbf{A}' - \lambda\mathbf{I}| = |\mathbf{A} - \lambda\mathbf{I}|$$

7-25

When a characteristic root, λ_i, is equal to zero, we have

$$\mathbf{A}\mathbf{x}_i = \lambda_i \mathbf{x}_i$$

$$= \mathbf{0}$$

From this equation, we find that there will be a nonzero characteristic vector $\mathbf{x}_i$, associated with a root λ_i equal to zero, only when the matrix **A** is singular.

Consider the matrix $\mathbf{A} = \begin{bmatrix} 1 & 0 \\ 2 & 0 \end{bmatrix}$ and determine the vector associated with the root $\lambda = 0$.

$$\det(\mathbf{A} - \lambda\mathbf{I}) = \begin{vmatrix} (1-\lambda) & 0 \\ 2 & -\lambda \end{vmatrix} = \lambda(\lambda - 1) = 0.$$

For $\lambda_1, = 0$ we have

$$(\mathbf{A} - \lambda_1\mathbf{I})\mathbf{x}_1 = \mathbf{0}$$

or,

$$\begin{bmatrix} 1 & 0 \\ 2 & 0 \end{bmatrix} \begin{bmatrix} x_1 \\ x_2 \end{bmatrix} = \begin{bmatrix} 0 \\ 0 \end{bmatrix}$$

Since det $\mathbf{A} = 0$, we can obtain a nonzero vector $\mathbf{x}_1$ such that

$$\mathbf{x}_1 = \beta \begin{bmatrix} 0 \\ 1 \end{bmatrix}$$

where β = arbitrary quantity.

7-26

Consider the matrix

$$\mathbf{A} = \begin{bmatrix} 10 & -2 & 4 \\ -20 & 4 & -10 \\ -30 & 6 & -13 \end{bmatrix}$$

where $\det(\mathbf{A} - \lambda\mathbf{I}) = (\lambda - 2)(\lambda + 1)(\lambda)$.

For the third root, $\lambda_3 = 0$, obtain the corresponding characteristic vector

$$\mathbf{x}_3 = \begin{bmatrix} \\ \\ \\ \end{bmatrix}$$

$$(\mathbf{A} - \lambda_3\mathbf{I})\mathbf{x}_3 = \mathbf{0}$$

yields

$$\begin{bmatrix} 10 & -2 & 4 \\ -20 & 4 & -10 \\ -30 & 6 & -13 \end{bmatrix} \begin{bmatrix} x_1 \\ x_2 \\ x_3 \end{bmatrix} = \begin{bmatrix} 0 \\ 0 \\ 0 \end{bmatrix}$$

$$\det \mathbf{A} = 10(-52 + 60) + 2(260 - 300) + 4(-120 + 120) = 0$$

or, **A** is singular and there is a nonzero vector, $\mathbf{x}_3$, which can be obtained. Adding twice the first row to the second row, we have

$$0 \cdot \mathbf{x}_1 + 0 \cdot \mathbf{x}_2 - 2\mathbf{x}_3 = 0$$

implying that $x_3 = 0$ and that x_1 can be set arbitrarily. The first row now yields

$$10x_1 - 2x_2 = 0$$

when $x_3 = 0$. Therefore, $x_2 = 5x_1$ and, letting $x_1 = \gamma$, we have

$$\mathbf{x}_3 = \gamma \begin{bmatrix} 1 \\ 5 \\ 0 \end{bmatrix}$$

7-27

If **B** is written as a scalar matrix, $\mathbf{B} = k\mathbf{A}$, then the equation

$$\mathbf{Bx} = \lambda\mathbf{x}$$

becomes

$$k\mathbf{Ax} = k\lambda\mathbf{x}$$

Hence, if a matrix **A** is multiplied by a constant, the characteristic roots of **A** are ______________________________ .

If a matrix **A** is multiplied by k, the characteristic roots of **A** are multiplied by the same constant k.

7-28

Recall that the characteristic roots and corresponding vectors of

$$\mathbf{A} = \begin{bmatrix} 1 & 2 \\ 2 & 1 \end{bmatrix}$$

are

$$\lambda_1 = -1, \mathbf{x}_1 = \beta \begin{bmatrix} 1 \\ -1 \end{bmatrix}$$

and

$$\lambda_2 = 3, \mathbf{x}_2 = \gamma \begin{bmatrix} 1 \\ 1 \end{bmatrix}$$

Determine the characteristic roots and corresponding vectors for

$$\mathbf{B} = \begin{bmatrix} 5 & 10 \\ 10 & 5 \end{bmatrix} = 5\mathbf{A}$$

$\lambda_1 =$ ____, $\mathbf{x}_1 = \beta \begin{bmatrix} \quad \\ \quad \end{bmatrix}$; $\lambda_2 =$ ____, $\mathbf{x}_2 = \gamma \begin{bmatrix} \quad \\ \quad \end{bmatrix}$

$\lambda_1 = -5, \mathbf{x}_1 = \beta \begin{bmatrix} 1 \\ -1 \end{bmatrix}$; $\lambda_2 = 15, \mathbf{x}_2 = \gamma \begin{bmatrix} 1 \\ 1 \end{bmatrix}$

If you answered correctly, proceed to Frame 7-29; if not, read the following explanation.

$$\lambda_1 = 5(-1) = -5, \lambda_2 = 5(3) = 15$$

Also, checking the roots we obtain

$$\det(\mathbf{B} - \lambda\mathbf{I}) = \det \begin{bmatrix} (5-\lambda) & 10 \\ 10 & (5-\lambda) \end{bmatrix} = \lambda^2 - 10\lambda - 75$$

$$= (\lambda + 5)(\lambda - 15)$$

for $\lambda_1 = -5$.

$$(\mathbf{B} - (-5)\mathbf{I})\mathbf{x}_1 = \mathbf{0}$$

or

$$\begin{bmatrix} 10 & 10 \\ 10 & 10 \end{bmatrix} \begin{bmatrix} x_1 \\ x_2 \end{bmatrix} = \begin{bmatrix} 0 \\ 0 \end{bmatrix}$$

which implies that

$$x_2 = -x_1$$

or

$$\mathbf{x}_1 = \beta \begin{bmatrix} 1 \\ -1 \end{bmatrix}$$

Similarly, for $\lambda_2 = 15$,

$$(\mathbf{B} - (15)\mathbf{I})\mathbf{x}_2$$

leads to

$$\begin{bmatrix} -10 & 10 \\ 10 & -10 \end{bmatrix} \mathbf{x}_2 = \mathbf{0}$$

which yields

$$x_1 = x_2$$

or

$$\mathbf{x}_2 = \gamma \begin{bmatrix} 1 \\ 1 \end{bmatrix}$$

7-29

We have found that if a matrix is multiplied by a constant k, the characteristic roots are multiplied by the constant k, but the characteristic vectors are not changed. In general, if we have a matrix

$$\mathbf{B} = k\mathbf{A}.$$

we can state that the characteristic roots of **B** are $k\lambda_i$ where λ_i are the roots of **A** and the *characteristic vectors of* **B** *are equal to the characteristic vectors of* **A**. If we are able to remove a scalar from the matrix **B**, the task of evaluating the characteristic roots and vectors will be greatly eased.

The characteristic roots of the inverse of a matrix $\mathbf{A}^{-1}$ are the inverses (reciprocals) of the characteristic roots of **A**, and both **A** and $\mathbf{A}^{-1}$ have the same characteristic vectors.

To prove this, consider the equation

$$\mathbf{Ax} = \lambda\mathbf{x}$$

and premultiply by $\mathbf{A}^{-1}$. Then as a second step, divide by λ to obtain:

$$\mathbf{A}^{-1}\mathbf{x} = \underline{\hspace{5cm}}$$

$$A^{-1}x = \frac{1}{\lambda}x$$

compared with $Ax = \lambda x$. Therefore, the characteristic roots of A^{-1} are the reciprocals $\frac{1}{\lambda_i}$ of the roots of **A**, whereas the characteristic vectors are the same.

7-30

For review, complete the following statements:

1. When a characteristic root is equal to zero, the vector associated with that root is equal to a zero vector when the matrix **A** is ________________ .
singular/nonsingular

2. If for a matrix, $B = kA$, and the characteristic roots and vectors of **A** are known, then the characteristic roots of **B** are______________ the character-
equal to/k times
istics of roots **A** and the characteristic vectors of **B** are ________________
equal to/k times
the characteristic vectors of **A**.

1. When a characteristic root is equal to zero, the vector associated with the root is equal to a zero vector when the matrix **A** is nonsingular. (see page 171 for review).

2. For a matrix $B = kA$ and the characteristic roots and vectors of **A** are known, then the characteristics root of **B** are k times the characteristics of roots **A**, and the characteristic vectors of **B** are equal to the characteristic vectors of **A**. (see page 177 for review).

This may be a convenient place for you to stop. Proceed when you are ready.

7-31

Recall that, for the orthogonal matrix, we have

$$\mathbf{A}^{-1} = \mathbf{A}'$$

The characteristic roots of $\mathbf{A}^{-1}$ are the reciprocals of the characteristic roots of $\mathbf{A}$, and the characteristic roots of $\mathbf{A}'$ are equal to the characteristic roots of $\mathbf{A}$. Therefore, for an *orthogonal matrix*, we have

$$\frac{1}{\lambda_i} = \lambda_i$$

or

$$\lambda_i^2 = 1$$

Hence, every characteristic root of an orthogonal matrix is either +1 or −1.

Find the characteristic roots and characteristic vectors for the orthogonal matrix

$$\mathbf{A} = \begin{bmatrix} \frac{1}{\sqrt{2}} & \frac{1}{\sqrt{2}} \\ \frac{1}{\sqrt{2}} & \frac{-1}{\sqrt{2}} \end{bmatrix}$$

$\lambda_1 =$ __________, $\mathbf{x}_1 = \beta\begin{bmatrix} \quad \\ \quad \end{bmatrix}$; $\lambda_2 =$ __________, $\mathbf{x}_2 = \gamma\begin{bmatrix} \quad \\ \quad \end{bmatrix}$

$$\lambda_1 = +1, \; \mathbf{x}_1 = \beta \begin{bmatrix} 1.0 \\ 0.414 \end{bmatrix}; \; \lambda_2 = -1, \; \mathbf{x}_2 = \gamma \begin{bmatrix} 1.0 \\ -2.414 \end{bmatrix}$$

If you had an error, read on; otherwise, proceed to Frame 7-32.

$$\det(\mathbf{A} - \lambda\mathbf{I}) = \det \begin{bmatrix} \left(\frac{1}{\sqrt{2}} - \lambda\right) & \frac{1}{\sqrt{2}} \\ \frac{1}{\sqrt{2}} & \left(\frac{-1}{\sqrt{2}} - \lambda\right) \end{bmatrix} = \lambda^2 - 1 = 0$$

$$\lambda = \pm 1.$$

Let $\lambda_1 = +1, \; \lambda_2 = -1$

When $\lambda = \lambda_1 = 1$,

$$(\mathbf{A} - \lambda_1\mathbf{I})\mathbf{x}_1 = \mathbf{0} \text{ and leads to}$$

$$\begin{bmatrix} \left(\frac{1}{\sqrt{2}} - 1\right) & \frac{1}{\sqrt{2}} \\ \frac{1}{\sqrt{2}} & \left(\frac{-1}{\sqrt{2}} - 1\right) \end{bmatrix} \begin{bmatrix} x_1 \\ x_2 \end{bmatrix} = \begin{bmatrix} 0 \\ 0 \end{bmatrix}$$

Row 1 implies that $\left(\frac{1}{\sqrt{2}} - 1\right) x_1 + \frac{1}{\sqrt{2}} x_2 = 0$

or

$$+0.414x_1 = +x_2$$

Therefore,

$$\mathbf{x}_1 = \beta \begin{bmatrix} 1.0 \\ 0.414 \end{bmatrix}$$

When $\lambda_2 = -1$, we obtain

$$\begin{bmatrix} \left(\frac{1}{\sqrt{2}}+1\right) & \frac{1}{\sqrt{2}} \\ \frac{1}{\sqrt{2}} & \left(\frac{-1}{\sqrt{2}}+1\right) \end{bmatrix} \begin{bmatrix} x_1 \\ x_2 \end{bmatrix} = \begin{bmatrix} 0 \\ 0 \end{bmatrix}$$

The first row implies that $\left(\frac{1}{\sqrt{2}} + 1\right)x_1 + \frac{1}{\sqrt{2}}x_2 = 0,$

or

$$2.414x_1 = -x_2$$

Therefore,

$$\mathbf{x}_2 = \gamma \begin{bmatrix} 1. \\ -2.414 \end{bmatrix}$$

7-32

The n characteristic roots of a triangular matrix or a diagonal matrix are the n diagonal elements of the matrix. First, consider the diagonal matrix **D**, which is written as

$$\mathbf{D} = \begin{bmatrix} d_{11} & 0 & \cdots & 0 \\ 0 & d_{22} & \cdots & 0 \\ \cdot & \cdot & & \cdot \\ \cdot & \cdot & & \cdot \\ \cdot & \cdot & & \cdot \\ 0 & 0 & \cdots & d_{nn} \end{bmatrix}$$

Then

$$\det(\mathbf{D} - \lambda\mathbf{I}) = (d_{11} - \lambda)(d_{22} - \lambda) \ldots (d_{nn} - \lambda) = 0$$

and clearly, the characteristic roots are equal to the diagonal elements of the matrix **D**.

Find the characteristic roots and the characteristic vectors for the matrix

$$\mathbf{F} = \begin{bmatrix} 1 & 0 \\ 0 & 2 \end{bmatrix}$$

$\lambda_1 =$ _____ $\qquad \mathbf{x}_1 = \qquad \mathbf{x}_2 =$

$\lambda_2 =$ _____

$\lambda_1 = 1 \qquad \mathbf{x}_1 = \begin{bmatrix} \alpha \\ 0 \end{bmatrix}$

$\lambda_2 = 2 \qquad \mathbf{x}_2 = \begin{bmatrix} 0 \\ \beta \end{bmatrix}$

If you have any questions about your answer read the following explanation, Otherwise proceed to Frame 7-33.

The two roots are the diagonal elements of the diagonal matrix. Therefore,

$$\lambda_1 = 1 \text{ and } \lambda_2 = 2$$

For $\lambda_1 = 1$ we have

$$(\mathbf{F} - \lambda_1 \mathbf{I})\mathbf{x}_1 = \mathbf{0}$$

$$\left(\begin{bmatrix} 1 & 0 \\ 0 & 2 \end{bmatrix} - 1 \begin{bmatrix} 1 & 0 \\ 0 & 1 \end{bmatrix}\right) \mathbf{x}_1 = \mathbf{0}$$

or

$$\begin{bmatrix} 0 & 0 \\ 0 & 1 \end{bmatrix} \begin{bmatrix} x_1 \\ x_2 \end{bmatrix} = \begin{bmatrix} 0 \\ 0 \end{bmatrix}$$

The second row requires that $x_2 = 0$ and x_1 is an arbitrary constant, α in this case, so that

$$\mathbf{x}_1 = \begin{bmatrix} \alpha \\ 0 \end{bmatrix}$$

For $\lambda_2 = 2$, we have

$$(\mathbf{F} - \lambda_2 \mathbf{I})\mathbf{x}_2 = \mathbf{0}$$

$$\left(\begin{bmatrix} 1 & 0 \\ 0 & 2 \end{bmatrix} - 2 \begin{bmatrix} 1 & 0 \\ 0 & 1 \end{bmatrix} \right) \mathbf{x}_2 = \mathbf{0}$$

$$\begin{bmatrix} -1 & 0 \\ 0 & 0 \end{bmatrix} \mathbf{x}_2 = \mathbf{0}$$

Therefore, it is necessary that $x_1 = 0$ and x_2 is arbitrary so that we will set it equal to β obtaining

$$\mathbf{x}_2 = \begin{bmatrix} 0 \\ \beta \end{bmatrix}$$

7-33

Show that the characteristic roots of the lower triangular matrix **B** are the diagonal elements where

$$\mathbf{B} = \begin{bmatrix} b_{11} & 0 & 0 \\ b_{21} & b_{22} & 0 \\ b_{31} & b_{32} & b_{33} \end{bmatrix}$$

$$\det(\mathbf{B} - \lambda\mathbf{I}) = \begin{vmatrix} (b_{11}-\lambda) & 0 & 0 \\ b_{21} & (b_{22}-\lambda) & 0 \\ b_{31} & b_{32} & (b_{33}-\lambda) \end{vmatrix}$$

Expanding along the first row, we obtain

$$\det(\mathbf{B} - \lambda\mathbf{I}) = (b_{11} - \lambda) \begin{vmatrix} (b_{22}-\lambda) & 0 \\ b_{32} & (b_{33}-\lambda) \end{vmatrix}$$

$$= (b_{11} - \lambda)(b_{22} - \lambda)(b_{33} - \lambda) = 0$$

Therefore, the characteristic roots of the triangular matrix **B** are equal to the elements on the diagonal. This is true for a triangular matrix of any order.

7-34

The sum of the characteristic roots of a matrix **A** is equal to the trace of **A**. Therefore,

$$\text{tr}(\mathbf{A}) = \sum_{i=1}^{n} \lambda_i$$

For example, when

$$\mathbf{A} = \begin{bmatrix} 1 & 1 & -2 \\ -1 & 2 & 1 \\ 0 & 1 & -1 \end{bmatrix},$$

we found $\lambda_1 = -1, \lambda_2 = 2, \lambda_3 = 1$

$$\text{tr}(\mathbf{A}) = 1 + 2 - 1 = 2$$

and

$$\sum_{i=1}^{3} \lambda_i = -1 + 2 + 1 = 2$$

SUMMARY

1. Given a square matrix **A** of order n. The characteristic value problem is to determine scalars λ and the associated nonzero vectors that simultaneously satisfy the equation

$$\mathbf{Ax} = \lambda\mathbf{x}$$

2. The polynomial equation in λ resulting from

$$\det(\mathbf{A} - \lambda\mathbf{I})$$

is called the characteristic polynomial.

3. The equation

$$\det(\mathbf{A} - \lambda\mathbf{I}) = 0$$

is called the characteristic equation, its roots are called the characteristic roots. The characteristic equation for a matrix of order n is an equation of order n with n roots.

4. Corresponding to each root λ_i of the characteristic equation, there is a characteristic vector $\mathbf{x}_i$ obtained from the homogeneous equation

$$(\mathbf{A} - \lambda_i\mathbf{I})\mathbf{x}_i = \mathbf{0}$$

5. The vectors $\mathbf{x}_1, \mathbf{x}_2, \ldots, \mathbf{x}_m$ are said to be linearly dependent if there exists a relationship among the vectors such that

$$k_1\mathbf{x}_1 + k_2\mathbf{x}_2 + \ldots + k_m\mathbf{x}_m = \mathbf{0}$$

Where the k_i are scalar numbers and at least one k_i is not zero. If no such relationship exists, the vectors are linearly independent.

6. The vectors $\mathbf{x}_1, \mathbf{x}_2, \ldots, \mathbf{x}_m$ are linearly dependent if and only if the rank of the matrix $\mathbf{M} = [\mathbf{x}_1, \mathbf{x}_2, \ldots, x_m]$ with the given vectors as columns less than m. They are independent if and only if the rank of **M** is equal to m.

7. A square matrix **B** is singular if and only if its columns (or rows) are linearly dependent.

8. When the characteristic roots of a matrix **A** are distinct and the associated characteristic vectors are nonzero, the vectors $\mathbf{x}_1, \mathbf{x}_2, ..., \mathbf{x}_n$ associated with the n characteristic roots are linearly independent. If there are repeated roots, the characteristic vectors may or may not be linearly independent.

9. There will be a nonzero characteristic vector $\mathbf{x}_i$ associated with a root λ_i that is equal to zero only when the matrix **A** is singular.

10. If a matrix is multiplied by scalar constant k so that

$$\mathbf{B} = k\mathbf{A}$$

then the characteristic roots of **B** are equal to the roots of **A** multiplied by k. The characteristic vectors of **B** are equal to the characteristic vectors of **A**.

11. The characteristic roots of the inverse of a matrix $\mathbf{A}^{-1}$ are the inverses (reciprocals) of the characteristic roots of **A**, and both **A** and $\mathbf{A}^{-1}$ have the same characteristic vectors.

12. Every characteristic root of an orthogonal matrix is either +1 or −1 and

$$\lambda_i^2 = 1$$

13. The n characteristic roots of a triangular matrix or a diagonal matrix are the n diagonal elements of the matrix.

14. The sum of the characteristic roots of a matrix **A** are equal to the trace of **A**. Therefore,

$$\text{tr}(\mathbf{A}) = \sum_{i=1}^{n} \lambda_i$$

15. The characteristic roots of a real symmetric matrix are all real.

EXERCISES

1. The characteristic roots of

$$\mathbf{A} = \begin{bmatrix} 1 & 0 & -1 \\ 1 & 2 & 1 \\ 2 & 2 & 3 \end{bmatrix}$$

are $\lambda_1 = 1, \lambda_2 = 2$, and $\lambda_3 = 3$.

Find the characteristic roots of $\mathbf{B} = 12\mathbf{A}$

$\lambda_1 =$ __________, $\lambda_2 =$ __________, $\lambda_3 =$ __________

2. Find the characteristic roots of the inverse of the matrix **A** in problem 1.

$\lambda_1 =$ __________, $\lambda_2 =$ __________, $\lambda_3 =$ __________

3. The characteristic roots of an orthogonal matrix are equal to __________ .

4. The sum of the characteristic roots of **B** is equal to __________ when

$$\mathbf{B} = \begin{bmatrix} 8 & 3 & 1 \\ 0 & 4 & 2 \\ 1 & 0 & 3 \end{bmatrix}$$

5. Determine the characteristic roots of **P** where

$$\mathbf{P} = \begin{bmatrix} 4 & 1 & 0 \\ 0 & -3 & 6 \\ 0 & 0 & 2 \end{bmatrix}$$

$\lambda_1 =$ __________, $\lambda_2 =$ __________, $\lambda_3 =$ __________

6. Consider three vectors

$$\mathbf{z}_1 = \begin{bmatrix} 4 \\ 2 \\ 0 \end{bmatrix}, \quad \mathbf{z}_2 = \begin{bmatrix} 2 \\ 1 \\ -3 \end{bmatrix}, \quad \mathbf{z}_3 = \begin{bmatrix} 6 \\ 3 \\ -5 \end{bmatrix}$$

Show that the three vectors are dependent.

7. Determine the characteristic roots and vectors for the matrix

$$\mathbf{Q} = \begin{bmatrix} 2 & 1 & 1 \\ 1 & 2 & 1 \\ 0 & 0 & 1 \end{bmatrix}$$

Are the characteristic vectors linearly independent?

8. Determine the characteristic vectors associated with the root $\lambda_1 = 0$ for the matrix

$$N = \begin{bmatrix} -2 & -8 & -12 \\ 1 & 4 & 4 \\ 0 & 0 & 1 \end{bmatrix}$$

$$x_1 = \beta \begin{bmatrix} \quad \end{bmatrix}$$

9. The characteristic roots of a real symmetric matrix are all ______________.

See page 246 for the answers.

8 matrix transformations and functions of a matrix

In analyzing problems in the social and physical sciences and in engineering, it is common to use two or more systems of coordinates. The *transformation* of a vector in one set of coordinates into a vector in another set is represented by

$$\mathbf{x} = \mathbf{T}\mathbf{y} \tag{8-1}$$

Where **T** is a *transformation matrix*. Thus we are interested in converting (or transforming) one vector into another more useful one. For example, let us consider two-element vectors **x** and **y** so that

$$\mathbf{x} = \begin{bmatrix} x_1 \\ \\ x_2 \end{bmatrix} \qquad \text{and } \mathbf{y} = \begin{bmatrix} y_1 \\ \\ y_2 \end{bmatrix}$$

The vector **y** is shown as $\mathbf{y} = \begin{bmatrix} 1 \\ 1 \end{bmatrix}$ in the figure

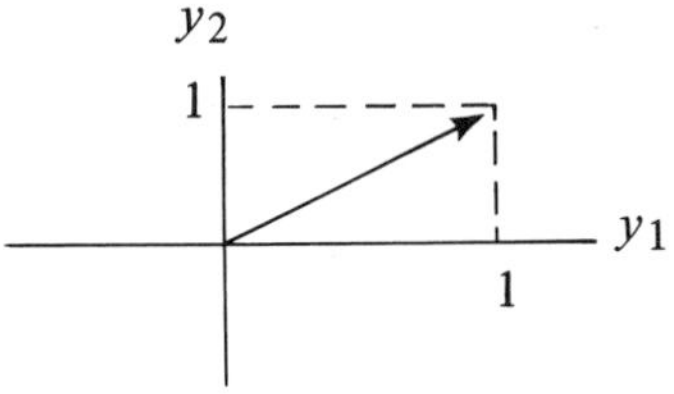

The vector **x** also appears on a plane with perpendicular coordinates x_1 and x_2.

Draw the coordinate system for **x**.

The **x** coordinate system appears as

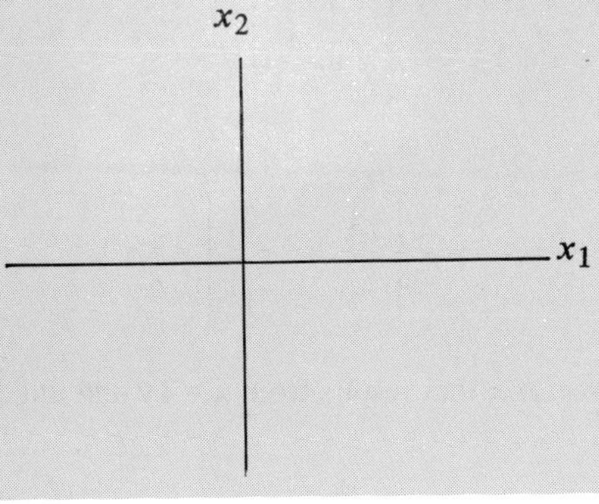

8-2

Equation (8-1) implies that a linear matrix transformation of the vector **y** into the vector **x** is possible through multiplication of the vector **y** by **T** yielding **x**.

Since

$$\mathbf{y} = \begin{bmatrix} 1 \\ 1 \end{bmatrix}$$

when

$$\mathbf{T} = \begin{bmatrix} -1 & 0 \\ 1 & 1 \end{bmatrix}$$

determine the vector **x** that results from **x** = **Ty** and indicate its location on the **x** plane.

$\mathbf{x} = \mathbf{Ty} =$

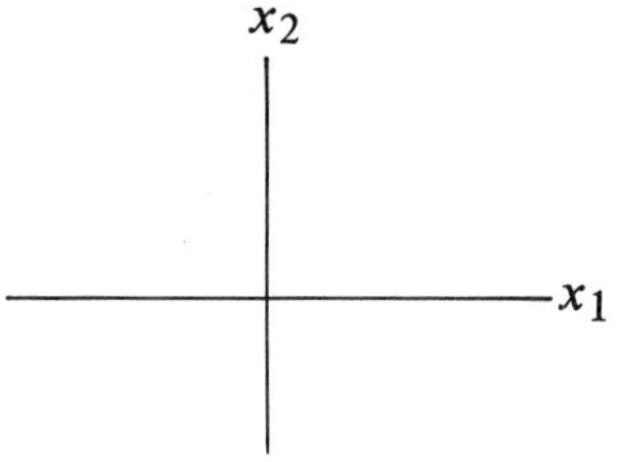

$$\mathbf{x} = \begin{bmatrix} -1 \\ 2 \end{bmatrix}$$

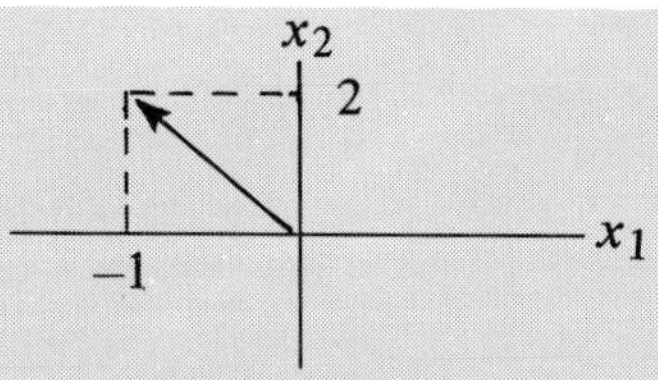

since

$$\mathbf{x} = \mathbf{Ty} = \begin{bmatrix} -1 & 0 \\ 1 & 1 \end{bmatrix} \begin{bmatrix} 1 \\ 1 \end{bmatrix} = \begin{bmatrix} -1 \\ 2 \end{bmatrix}$$

8-3

The multiplication by the matrix **T** converts (transforms) one vector, **y** into another vector, **x**.

The transformation is

$$\mathbf{x} = \mathbf{T}\mathbf{y} \qquad (8\text{-}1)$$

If **T** is nonsingular we can premultiply both sides of equation (8-1) by $\mathbf{T}^{-1}$ to obtain

y = ____________________

$$\mathbf{y} = \mathbf{T}^{-1}\mathbf{x} \qquad (8\text{-}2)$$

Thus, **x** is retransformed into **y** by the inverse of matrix $\mathbf{T}^{-1}$.

8-4

For

$$\mathbf{x} = \begin{bmatrix} -1 \\ 2 \end{bmatrix} \text{ and } \mathbf{T} = \begin{bmatrix} -1 & 0 \\ 1 & 1 \end{bmatrix}$$

find $\mathbf{T}^{-1}$ and show that

$$\mathbf{y} = \mathbf{T}^{-1}\mathbf{x} = \begin{bmatrix} 1 \\ 1 \end{bmatrix}$$

$$\mathbf{T}^{-1} = \begin{bmatrix} -1 & 0 \\ 1 & 1 \end{bmatrix} \quad \text{since } \det \mathbf{T} = -1.$$

and

$$\mathbf{y} = \mathbf{T}^{-1}\mathbf{x} = \begin{bmatrix} -1 & 0 \\ 1 & 1 \end{bmatrix} \begin{bmatrix} -1 \\ 2 \end{bmatrix} = \begin{bmatrix} 1 \\ 1 \end{bmatrix}$$

8-5

The n characteristic vectors of a matrix **A** can be arranged as the columns of a square matrix **M**, which is called the *modal matrix*. Therefore,

$$\mathbf{M} = [\mathbf{x}_1, \mathbf{x}_2, ..., \mathbf{x}_n] \tag{8-3}$$

when $\mathbf{x}_i$ is the ith characteristic vector. The modal matrix is useful for transforming from one set of coordinates into another more useful form. There are an infinite number of modal matrices because a characteristic vector is multiplied by an arbitrary scalar β, γ, or a similar constant. We will adopt the convention of setting the scalar constant equal to one in all cases.

Consider the three characteristic vectors obtained on pages 158-160 for the mat

$$\mathbf{A} = \begin{bmatrix} 1 & 1 & -2 \\ -1 & 2 & 1 \\ 0 & 1 & -1 \end{bmatrix}$$

$$\mathbf{x}_1 = \beta \begin{bmatrix} 1 \\ 0 \\ 1 \end{bmatrix} \quad \mathbf{x}_2 = \gamma \begin{bmatrix} 1 \\ 3 \\ 1 \end{bmatrix} \quad \mathbf{x}_3 = \delta \begin{bmatrix} 3 \\ 2 \\ 1 \end{bmatrix}$$

Set $\beta = \gamma = \delta = 1$ and determine the modal matrix.

$$\mathbf{M} = \begin{bmatrix} \quad \end{bmatrix}$$

$$\mathbf{M} = \begin{bmatrix} 1 & 1 & 3 \\ 0 & 3 & 2 \\ 1 & 1 & 1 \end{bmatrix}$$

8-6

A matrix related to the characteristic roots of the matrix **A** is the *spectral matrix* **S**, which is a diagonal matrix with the characteristic roots along the major diagonal so that

$$\mathbf{S} = \begin{bmatrix} \lambda_1 & 0 \ldots & 0 \\ 0 & \lambda_2 \ldots & 0 \\ \vdots & \vdots & \vdots \\ 0 & 0 & \lambda_n \end{bmatrix} \tag{8-4}$$

For example, for the matrix

$$A = \begin{bmatrix} 1 & 1 & -2 \\ -1 & 2 & 1 \\ 0 & 1 & -1 \end{bmatrix}$$

we found that $\lambda_1 = -1$, $\lambda_2 = 2$, and $\lambda_3 = 1$.

Complete the spectral matrix for this set of characteristic roots.

$$S = \begin{bmatrix} -1 & 0 & 0 \\ \_\_ & \_\_ & \_\_ \\ 0 & 0 & \_\_ \end{bmatrix}$$

The characteristic roots of the spectral matrix are ____________________.

$$S = \begin{bmatrix} -1 & 0 & 0 \\ 0 & 2 & 0 \\ 0 & 0 & 1 \end{bmatrix}$$

The characteristic roots of the spectral matrix are $-1, 2, 1$ which are the same as the characteristic roots of **A**. Recall that the characteristic roots of a diagonal matrix are equal to the elements of the diagonal.

8-7

The characteristic roots of the spectral matrix and the characteristic roots of the original matrix **A** are the same.

Let's review the concept of a modal and spectral matrix. The modal matrix **M** consists of the ____________________ characteristic vectors associated with a matrix **A** of order n. The modal matrix is useful in ____________________ from one set of coordinates into another. If a matrix **A** is of order 3 and has three distinct characteristic roots, write the spectral matrix **S**.

The modal matrix **M** consists of the n characteristic vectors associated with a matrix **A** of order n. The modal matrix is useful in transforming (converting) from one set of coordinates into another.

$$\mathbf{S} = \begin{bmatrix} \lambda_1 & 0 & 0 \\ 0 & \lambda_2 & 0 \\ 0 & 0 & \lambda_3 \end{bmatrix}$$

8-8

We are interested in obtaining the matrix relationship between the original matrix **A** and the spectral matrix **S**. The characteristic vectors $\mathbf{x}_i$ contained in the modal matrix satisfy the equation $\mathbf{A}\mathbf{x}_i = \lambda_i \mathbf{x}_i$

Recall that $\mathbf{M} = [\mathbf{x}_1, \mathbf{x}_2, ..., \mathbf{x}_n]$ and premultiply **M** by **A** to obtain

$$\mathbf{AM} = \mathbf{A}[\mathbf{x}_1, \mathbf{x}_2, ..., \mathbf{x}_n] \qquad (8\text{-}5)$$
$$= [\mathbf{A}\mathbf{x}_1, \mathbf{A}\mathbf{x}_2, ..., \mathbf{A}\mathbf{x}_n]$$

Utilizing the fact that $\mathbf{A}\mathbf{x}_i = \lambda_i \mathbf{x}_i$, we have

$$\mathbf{AM} = \begin{bmatrix} \qquad\qquad \end{bmatrix}$$

$$\mathbf{AM} = [\lambda_1 \mathbf{x}_1, \lambda_2 \mathbf{x}_2, \ldots, \lambda_n \mathbf{x}_n]$$

Since

$$\mathbf{S} = \begin{bmatrix} \lambda_1 & 0 & \ldots & 0 \\ 0 & \lambda_2 & \ldots & 0 \\ \vdots & \vdots & & \vdots \\ 0 & 0 & \ldots & \lambda_n \end{bmatrix}$$

we recognize that

$$[\lambda_1 \mathbf{x}_1, \lambda_2 \mathbf{x}_2, \ldots, \lambda_n \mathbf{x}_n] = \mathbf{MS}$$

and, therefore,

(8-6) $$\mathbf{AM} = \mathbf{MS}$$

Premultiplying by $\mathbf{M}^{-1}$ we obtain the relationship between the spectral matrix **S** and the original matrix **A**, which is, in terms of the modal matrix **M**,

(8-7) $$\mathbf{M}^{-1}\mathbf{AM} = \mathbf{S}$$

8-9

For a matrix **A** of order 2 with two distinct characteristic roots, λ_1 and λ_2, complete the algebra utilized in proceeding from equation (8-5) to (8-6) in order to verify that equation (8-6) is a correct relationship.

$$\mathbf{M} = [\mathbf{x}_1, \mathbf{x}_2]$$

$$\mathbf{AM} = [\mathbf{Ax}_1, \mathbf{Ax}_2]$$

$$= [\lambda_1 \mathbf{x}_1, \lambda_2 \mathbf{x}_2]$$

$$= [\mathbf{x}_1, \mathbf{x}_2] \begin{bmatrix} \lambda_1 & 0 \\ 0 & \lambda_2 \end{bmatrix}$$

$$= \mathbf{MS}$$

Equation (8-7)

$$\mathbf{M}^{-1}\mathbf{A}\mathbf{M} = \mathbf{S}$$

represents the *diagonalization* of the matrix **A**, or the *transformation* of **A** into a diagonal matrix. A square matrix **A** of order *n* can be diagonalized if the matrix has *n* linearly independent characteristic vectors.

A matrix **A** always has *n* linearly independent vectors if it has *n* distinct characteristic roots or if it is a symmetric matrix. If the matrix has repeated characteristic roots, it may or may not have linearly independent characteristic vectors.

Consider the matrix

$$\mathbf{A} = \begin{bmatrix} 1 & 2 \\ 2 & 1 \end{bmatrix}$$

where

$$\lambda_1 = -1, \mathbf{x}_1 = \begin{bmatrix} 1 \\ -1 \end{bmatrix} \qquad \text{and } \lambda_2 = 3, \mathbf{x}_2 = \begin{bmatrix} 1 \\ 1 \end{bmatrix}$$

Find **M**, $\mathbf{M}^{-1}$, and then **S**.

$$\mathbf{M} = \begin{bmatrix} 1 & 1 \\ -1 & 1 \end{bmatrix}, \mathbf{M}^{-1} = \frac{1}{2}\begin{bmatrix} 1 & -1 \\ 1 & 1 \end{bmatrix}, \mathbf{S} = \begin{bmatrix} -1 & 0 \\ 0 & 3 \end{bmatrix}$$

Algebra;

$$\mathbf{S} = \mathbf{M}^{-1}\ \mathbf{AM} = \mathbf{M}^{-1}\begin{bmatrix} 1 & 2 \\ 2 & 1 \end{bmatrix}\begin{bmatrix} 1 & 1 \\ -1 & 1 \end{bmatrix}$$

$$= \frac{1}{2}\begin{bmatrix} 1 & -1 \\ 1 & 1 \end{bmatrix}\begin{bmatrix} -1 & 3 \\ 1 & 3 \end{bmatrix}$$

$$= \frac{1}{2}\begin{bmatrix} -2 & 0 \\ 0 & 6 \end{bmatrix}\begin{bmatrix} -1 & 0 \\ 0 & 3 \end{bmatrix}$$

8-11

In problem 7 of the exercises of Chapter 7 (page 188 we considered the matrix

$$\mathbf{Q} = \begin{bmatrix} 2 & 1 & 1 \\ 1 & 2 & 1 \\ 0 & 0 & 1 \end{bmatrix}$$

where

$$\lambda_1 = 1, \mathbf{x}_1 = \begin{bmatrix} 1 \\ 0 \\ -1 \end{bmatrix}, \lambda_2 = 1, \mathbf{x}_2 = \begin{bmatrix} 0 \\ 1 \\ -1 \end{bmatrix}, \lambda_3 = 3, \mathbf{x}_3 = \begin{bmatrix} 1 \\ 1 \\ 0 \end{bmatrix}$$

There is a repeated root, since

$$\lambda_1 = \lambda_2 = 1$$

Can a diagonal matrix **S** be obtained? ______ . Why?
Yes/No

If so, specify the matrices **M** and **S**.

$$\mathbf{M} = \begin{bmatrix} \quad & \quad & \quad \end{bmatrix}$$

$$\mathbf{S} = \begin{bmatrix} \quad & \quad & \quad \end{bmatrix}$$

Yes, a diagonal matrix **S** can be obtained because the characteristic vectors are independent.

$$\mathbf{M} = \begin{bmatrix} 1 & 0 & 1 \\ 0 & 1 & 1 \\ -1 & -1 & 0 \end{bmatrix},$$

det $\mathbf{M} \neq 0$, and, therefore, the vectors are independent, (Refer to Frame 7-19)

$$\mathbf{S} = \begin{bmatrix} 1 & 0 & 0 \\ 0 & 1 & 0 \\ 0 & 0 & 3 \end{bmatrix}$$

8-12

The two matrices, **S**, and **A**, are said to be *similar* is there exists a non-singular matrix **M** such that

$$\mathbf{S} = \mathbf{M}^{-1}\mathbf{AM}$$

As we have shown earlier in this chapter, two similar matrices possess the same characteristic roots.

An $n \times n$ symmetric matrix has the property that a diagonal matrix **S** can always be obtained by an appropriate similarity transformation **M**. This is because even if a symmetric matrix has repeated characteristic roots, the characteristic vectors of a symmetric matrix are always linearly independent. Also, the characteristic roots of a real symmetric matrix are always real.

Consider the symmetric matrix

$$\mathbf{A} = \begin{bmatrix} 7 & -2 & 1 \\ -2 & -10 & -2 \\ 1 & -2 & 7 \end{bmatrix}$$

where the characteristic roots are $\lambda_1 = \lambda_2 = 6$ and $\lambda_3 = 12$.

Determine the modal matrix and the spectral matrix.

$$\mathbf{M} = \begin{bmatrix} -1 & 1 & 1 \\ 0 & 1 & -2 \\ 1 & 1 & 1 \end{bmatrix} \quad \mathbf{S} = \begin{bmatrix} 6 & 0 & 0 \\ 0 & 6 & 0 \\ 0 & 0 & 12 \end{bmatrix}$$

If your answer was correct, proceed to Frame 8-13; otherwise, read below.

For $\lambda = 6$, we have $(\mathbf{A} - \lambda_1 \mathbf{I})\mathbf{x} = 0$

$$\begin{bmatrix} 1 & -2 & 1 \\ -2 & 4 & -2 \\ 1 & -2 & 1 \end{bmatrix} \begin{bmatrix} x_1 \\ x_2 \\ x_3 \end{bmatrix} = \mathbf{0} \quad \text{or} \quad x_1 - 2x_2 + x_3 = 0$$

For x_1 we will let $x_2 = 0$, so that $x_1 = -x_3$. Therefore,

$$\mathbf{x}_1 = \begin{bmatrix} -1 \\ 0 \\ 1 \end{bmatrix}$$

For $\mathbf{x}_2$ an alternate and independent solution to $x_1 - 2x_2 + x_3 = 0$ is

$$\mathbf{x}_2 = \begin{bmatrix} 1 \\ 1 \\ 1 \end{bmatrix}$$

Finally, for $\lambda_3 = 12$, we obtain

$$\mathbf{x}_3 = \begin{bmatrix} 1 \\ -2 \\ 1 \end{bmatrix}$$

Note that

$$\det \mathbf{M} = \begin{vmatrix} -1 & 1 & 1 \\ 0 & 1 & -2 \\ 1 & 1 & 1 \end{vmatrix} = -1 \begin{vmatrix} 1 & -2 \\ 1 & 1 \end{vmatrix} + 1 \begin{vmatrix} 1 & 1 \\ 1 & -2 \end{vmatrix}$$

$$= -1(3) + 1(-3) = -6$$

and the vectors are linearly independent, as is always the case for a symmetric matrix.

8-13

This is a good time to briefly review the last few concepts. Complete the following questions:

1. The equation $\mathbf{S}$ = ____________ represents the transformation of the matrix **A** into a diagonal matrix.

2. A matrix with repeated roots ____________ have linearly independent characteristic vectors.
   may/may not

3. The diagonal spectral matrix can be obtained for a square matrix **A** of order *n* when the characteristic vectors are linearly ____________________.
   independent/dependent/either

1. The equation $\mathbf{S} = \mathbf{M}^{-1}\mathbf{AM}$ represents the transformation of the matrix **A** into a diagonal matrix.

2. A matrix with repeated roots may have linearly independent characteristic vectors.

3. The diagonal spectral matrix can be obtained for a square matrix **A** of order *n* when the characteristic vectors are linearly independent.

8-14

If **A** is a symmetric matrix with real elements, the *normalized modal matrix* **N** associated with **A** is an orthogonal matrix and therefore,

$$\mathbf{N}' = \mathbf{N}^{-1}$$

(See Frame 6-13)

Therefore, we may write the equation for the spectral matrix as

$$\mathbf{S} = \mathbf{N}^{-1}\mathbf{A}\mathbf{N} = \mathbf{N}'\mathbf{A}\mathbf{N}$$

A modal matrix **M** is normalized by adjusting the arbitrary constant in each characteristic vector such that the sum of the square of the elements in the column is equal to 1.

For the matrix

$$\mathbf{A} = \begin{bmatrix} 7 & -2 & 1 \\ -2 & 10 & -2 \\ 1 & -2 & 7 \end{bmatrix}$$

we found on page 204, that

$$\mathbf{M} = \begin{bmatrix} -1 & 1 & 1 \\ 0 & 1 & -2 \\ 1 & 1 & 1 \end{bmatrix}$$

since

$$\mathbf{x}_1 = \beta\begin{bmatrix} -1 \\ 0 \\ 1 \end{bmatrix}, \mathbf{x}_2 = \gamma\begin{bmatrix} 1 \\ 1 \\ 1 \end{bmatrix}, \mathbf{x}_3 = \delta\begin{bmatrix} 1 \\ -2 \\ 1 \end{bmatrix}$$

Obtain **N**, show that it is an orthogonal matrix and that $\mathbf{S} = \mathbf{N}'\mathbf{A}\mathbf{N}$.

$$\mathbf{N} = \begin{bmatrix} \frac{-1}{\sqrt{2}} & \frac{1}{\sqrt{3}} & \frac{1}{\sqrt{6}} \\ 0 & \frac{1}{\sqrt{3}} & \frac{-2}{\sqrt{6}} \\ \frac{1}{\sqrt{2}} & \frac{1}{\sqrt{3}} & \frac{1}{\sqrt{6}} \end{bmatrix}$$

The matrix **N** is orthogonal since $\mathbf{N}^{-1} = \mathbf{N}'$ or $\mathbf{NN}' = \mathbf{I}$

$$\mathbf{NN}' = \begin{bmatrix} \frac{-1}{\sqrt{2}} & \frac{1}{\sqrt{3}} & \frac{1}{\sqrt{6}} \\ 0 & \frac{1}{\sqrt{3}} & \frac{-2}{\sqrt{6}} \\ \frac{1}{\sqrt{2}} & \frac{1}{\sqrt{3}} & \frac{1}{\sqrt{6}} \end{bmatrix} \begin{bmatrix} \frac{-1}{\sqrt{2}} & 0 & \frac{1}{\sqrt{2}} \\ \frac{1}{\sqrt{3}} & \frac{1}{\sqrt{3}} & \frac{1}{\sqrt{3}} \\ \frac{1}{\sqrt{6}} & \frac{-2}{\sqrt{6}} & \frac{1}{\sqrt{6}} \end{bmatrix} = \begin{bmatrix} 1 & 0 & 0 \\ 0 & 1 & 0 \\ 0 & 0 & 1 \end{bmatrix}$$

$$\mathbf{S} = \mathbf{N}'\mathbf{AN} = \begin{bmatrix} 6 & 0 & 0 \\ 0 & 6 & 0 \\ 0 & 0 & 12 \end{bmatrix}$$

8-15

Consider a polynomial of the algebraic variable of z of degree n where z may be a real or complex number, which may be written as

$$P(z) = c_0 + c_1 z + c_2 z^2 + \ldots + c_n z^n$$

The coefficients c_i are known constants. If we replace the scalar variable z with the matrix **A** we obtain the *matrix polynomial*

$$P(\mathbf{A}) = c_0\mathbf{I} + c_1\mathbf{A} + c_2\mathbf{A}^2 + \ldots + c_n\mathbf{A}^n$$

where **I** = identity matrix and $\mathbf{A}^k$ implies that **A** is multiplied by itself k times. $P(\mathbf{A})$ is defined as a matrix polynomial and is a matrix itself.

If

$$\mathbf{A} = \begin{bmatrix} 2 & 2 \\ 1 & 0 \end{bmatrix}$$

evaluate the matrix polynomial

$$P(\mathbf{A}) = \mathbf{I} + 2\mathbf{A} + 3\mathbf{A}^2$$

P(A) =

$$P(\mathbf{A}) = \begin{bmatrix} 23 & 16 \\ 8 & 7 \end{bmatrix}$$

since

$$P(\mathbf{A}) = \begin{bmatrix} 1 & 0 \\ 0 & 1 \end{bmatrix} + 2 \begin{bmatrix} 2 & 2 \\ 1 & 0 \end{bmatrix} + 3 \begin{bmatrix} 6 & 4 \\ 2 & 2 \end{bmatrix} = \begin{bmatrix} 23 & 16 \\ 8 & 7 \end{bmatrix}$$

where

$$\mathbf{A}^2 = \mathbf{A} \cdot \mathbf{A} = \begin{bmatrix} 6 & 4 \\ 2 & 2 \end{bmatrix}$$

8-16

Therefore, we may write a *function* of a *matrix* in the same manner that we write the function of a variable. One particularly valuable function is the *characteristic polynomial* of the matrix **A**, which is written as

$$f(\mathbf{A}) = c_0\mathbf{I} + c_1\mathbf{A} + c_2\mathbf{A}^2 + \ldots + c_n\mathbf{A}^n$$

where $\mathbf{A}^k$ implies that **A** is multiplied k times and c_i = ith scalar constant. These coefficients c_i are the coefficients obtained in the characteristic polynomial (see page 149) as

$$\det(\mathbf{A} - \lambda\mathbf{I}) = c_0 + c_1\lambda + c_2\lambda^2 + \ldots + c_n\lambda^n = f(\lambda)$$

The characteristic equation is $f(\lambda) = 0$. Write the characteristic equation for a matrix of order 2 in terms of the generalized scalar constants, c_i.

$f(\lambda) =$ ____________________

$f(\lambda) = c_0 + c_1\lambda + c_2\lambda^2 = 0$ is the characteristic equation for a matrix of order 2. (See page 149 for review).

Since we can write the function of a matrix $f(\mathbf{A})$, we expect that one can obtain a matrix equation similar to the scalar equation $f(\lambda) = 0$. In fact, a theorem provides just such a relation.

The *Cayley-Hamilton Theorem* states that a matrix satisfies its own characteristic equation; that is

$$f(\mathbf{A}) = \mathbf{0}$$

Consider the matrix

$$\mathbf{A} = \begin{bmatrix} 1 & 2 \\ 2 & 1 \end{bmatrix}$$

where det $(\mathbf{A}-\lambda\mathbf{I}) \;=\; \lambda^2 - 2\lambda - 3 = 0.$

Show that $f(\mathbf{A}) = \mathbf{0}$ and that the matrix **A** satisfies its own characteristic equation.

$$f(\mathbf{A}) = \mathbf{A}^2 - 2\mathbf{A} - 3\mathbf{I}$$

$$\mathbf{A}^2 = \begin{bmatrix} 1 & 2 \\ 2 & 1 \end{bmatrix}\begin{bmatrix} 1 & 2 \\ 2 & 1 \end{bmatrix} = \begin{bmatrix} 5 & 4 \\ 4 & 5 \end{bmatrix}, \quad 2\mathbf{A} = \begin{bmatrix} 2 & 4 \\ 4 & 2 \end{bmatrix}$$

and therefore

$$\mathbf{A}^2 - 2\mathbf{A} - 3\mathbf{I} = \begin{bmatrix} 5 & 4 \\ 4 & 5 \end{bmatrix} - \begin{bmatrix} 2 & 4 \\ 4 & 2 \end{bmatrix} - \begin{bmatrix} 3 & 0 \\ 0 & 3 \end{bmatrix} = \begin{bmatrix} 0 & 0 \\ 0 & 0 \end{bmatrix}$$

or $f(\mathbf{A}) = \mathbf{0}$ and the matrix satisfies its own characteristic equation.

An important application of the Cayley-Hamilton Theorem is in the representation of high powers of a matrix, as we will find on page 216.

8-18

An important and useful polynomial function of the scalar variable t is the *exponential function* $\exp(t) = e^t$. The exponential function is defined as the infinite series

$$\exp(t) = e^t = 1 + \frac{t}{1!} + \frac{t^2}{2!} + \ldots + \frac{t^k}{k!} + \ldots$$

$$= \sum_{k=0}^{\infty} \frac{t^k}{k!}$$

This series converges for all values of t.

Write the series in expanded form and in the summation form when $t = 1$.

$e^1 =$

$$e^1 = 1 + 1 + \frac{1}{2} + \frac{1}{3!} + \frac{1}{4!} + \ldots + \frac{1}{k!} + \ldots$$

$$= \sum_{k=0}^{\infty} \frac{1}{k!}$$

In calculus it is shown that e is the base of natural logarithms and the series e^1 is approximately equal to $e = 2.71828$. You can obtain this value for e by adding ten terms of the series.

8-19

A polynomial function of a matrix **A** of great utility is the *matrix exponential function,* which is defined as the power series

$$\exp(\mathbf{A}) = e^{\mathbf{A}} = \mathbf{I} + \frac{\mathbf{A}}{1!} + \frac{\mathbf{A}^2}{2!} + \ldots + \frac{\mathbf{A}^k}{k!} + \ldots$$

$$= \sum_{k=0}^{\infty} \frac{\mathbf{A}^k}{k!}$$

where $\mathbf{A}^k$ implies **A** multiplied by itself k times. This series may be shown to be convergent for all square matrices.

When

$$\mathbf{A} = \begin{bmatrix} 0 & 1 \\ 0 & 0 \end{bmatrix}$$

find the function exp (**A**).

$$\exp(\mathbf{A}) = \begin{bmatrix} 1 & 1 \\ 0 & 1 \end{bmatrix}$$

since

$$\mathbf{A}^2 = \begin{bmatrix} 0 & 1 \\ 0 & 0 \end{bmatrix} \begin{bmatrix} 0 & 1 \\ 0 & 0 \end{bmatrix} = \begin{bmatrix} 0 & 0 \\ 0 & 0 \end{bmatrix}$$

and

$$\mathbf{A}^k = \mathbf{0}, \text{ the null matrix, for } k \geqslant 2$$

Therefore,

$$\exp(\mathbf{A}) = \mathbf{I} + \mathbf{A} = \begin{bmatrix} 1 & 0 \\ 0 & 1 \end{bmatrix} + \begin{bmatrix} 0 & 1 \\ 0 & 0 \end{bmatrix}$$

$$= \begin{bmatrix} 1 & 1 \\ 0 & 1 \end{bmatrix}$$

Note that a function of a matrix is a matrix itself.

8-20

If **A** and **B** are commutative matrices (that is, **AB** = **BA**) then we have

$$\exp(\mathbf{A}) \cdot \exp(\mathbf{B}) = \exp(\mathbf{B}) \cdot \exp(\mathbf{A})$$

$$= \exp(\mathbf{A} + \mathbf{B})$$

It can also be shown that exp (−**A**) is the inverse of the matrix exp (**A**) or

$$(\exp(\mathbf{A}))^{-1} \cdot \exp(\mathbf{A}) = \exp(-\mathbf{A}) \cdot \exp(\mathbf{A}) = \mathbf{I}$$

If

$$\mathbf{A} = \begin{bmatrix} 0 & 1 \\ 0 & 0 \end{bmatrix} \text{ and } \exp(\mathbf{A}) = \begin{bmatrix} 1 & 1 \\ 0 & 1 \end{bmatrix}$$

Find $\exp(-\mathbf{A})$ and show that $\exp(-\mathbf{A}) = (\exp(\mathbf{A}))^{-1}$

$$\exp(-\mathbf{A}) = \exp\left(\begin{bmatrix} 0 & -1 \\ 0 & 0 \end{bmatrix}\right) = \mathbf{I} + (-\mathbf{A}) = \begin{bmatrix} 1 & -1 \\ 0 & 1 \end{bmatrix}$$

since

$$\mathbf{A}^k = \mathbf{0} \text{ for } k > 2$$

$$(\exp(\mathbf{A}))^{-1} = \begin{bmatrix} 1 & 1 \\ 0 & 1 \end{bmatrix}^{-1} = \begin{bmatrix} 1 & -1 \\ 0 & 1 \end{bmatrix}$$

and

$$\exp(-\mathbf{A}) = (\exp(\mathbf{A}))^{-1}$$

The exponential matrix is useful in the solution of matrix differential equations. This subject is beyond the scope of this book, but you might wish to refer to one of the following references:

1. L. Pipes, *Matrix Methods for Engineering,* Englewood Cliffs, N.J., Prentice-Hall, 1963.

2. R.C. Dorf, *Time-Domain Analysis and Design of Control Systems,* Reading, Addison-Wesley, 1965

8-21

Because a matrix satisfies its own characteristic equation in terms of the original matrix **A** we have

$$f(\mathbf{A}) = c_0\mathbf{I} + c_1\mathbf{A}^2 + \ldots + c_n\mathbf{A}^n = \mathbf{0}$$

Solving for the highest power $\mathbf{A}^n$ we have

$$\mathbf{A}^n = -(\frac{c_{n-1}}{c_n}\ \mathbf{A}^{n-1} + \frac{c_{n-2}}{c_n}\ \mathbf{A}^{n-2} + \ldots + \frac{c_0}{c_n}\mathbf{I})$$

and we can express the highest power of **A** in terms of the lower order terms.

Let us use this property of a matrix function to find the matrix polynomial

$$P(\mathbf{A}) = 3\mathbf{A} + \mathbf{A}^4$$

where

$$\mathbf{A} = \begin{bmatrix} 1 & 2 \\ 3 & 0 \end{bmatrix}$$

The characteristic equation is $f(\lambda) =$ __________ and the corresponding matrix equation is $f(\mathbf{A}) =$ ____________________.

The characteristic equation is $f(\lambda) = \lambda^2 - \lambda - 6 = 0$ and the corresponding matrix equation is

$$f(\mathbf{A}) = \mathbf{A}^2 - \mathbf{A} - 6\mathbf{I} = 0$$

The characteristic matrix equation can be written

$$\mathbf{A}^2 = \mathbf{A} + 6\mathbf{I}$$

and therefore,

$$\begin{aligned} \mathbf{A}^4 &= \mathbf{A}^2 \cdot \mathbf{A}^2 = \mathbf{A}^2 + 12\mathbf{A} + 36\mathbf{I} \\ &= (\mathbf{A} + 6\mathbf{I}) + 12\mathbf{A} + 36\mathbf{I} \\ &= 13\mathbf{A} + 42\mathbf{I} \end{aligned}$$

8-22

Therefore, the polynomial

$$f(\mathbf{A}) = 3\mathbf{A} + \mathbf{A}^4$$

written in terms of **A** to the first power and **I** is

$$f(\mathbf{A}) = \underline{\hspace{4cm}}$$

$$f(\mathbf{A}) = 16\mathbf{A} + 42\mathbf{I}$$

since

$$\mathbf{A}^4 = 13\mathbf{A} + 42\mathbf{I}$$

This is a suitable stopping place in Chapter 8. If you are familiar with the calculus you will profit from the next section, which is a study of the derivative of a matrix, which is useful in the calculus of matrices. If you wish to bypass the section, skip to the review section on page 221.

THE DERIVATIVE OF A MATRIX

8-23

In the calculus we are interested in the derivative of a function of the variable t, which is written

$$\frac{d}{dt}(f(t)) = \frac{df(t)}{dt}$$

Similarly, in the calculus of matrices we are interested in obtaining the derivative of a matrix

$$\mathbf{A}(t) = \begin{bmatrix} a_{11}(t) & \dots & a_{1n}(t) \\ \cdot & & \\ \cdot & & \\ \cdot & & \\ a_{n1}(t) & \dots & a_{nn}(t) \end{bmatrix}$$

whose elements are functions of the variable t.

The derivative of a matrix $\mathbf{A}(t)$, whose elements are functions of the variable t, is defined as

$$\frac{d}{dt}(\mathbf{A}(t)) = \begin{bmatrix} \frac{da_{11}(t)}{dt} & \dots & \frac{da_{1n}(t)}{dt} \\ \cdot & & \\ \cdot & & \\ \cdot & & \\ \frac{da_{n1}(t)}{dt} & \dots & \frac{da_{nn}(t)}{dt} \end{bmatrix}$$

That is, the derivative of a matrix is simply the derivative of each element $a_{ij}(t)$ of the matrix.

Obtain $\frac{d}{dt}(\mathbf{A}(t))$ when

$$\mathbf{A}(t) = \begin{bmatrix} e^{-2t} & 3t \\ t^2 & 2 \end{bmatrix}, \quad \frac{d\mathbf{A}(t)}{dt} = \begin{bmatrix} & \\ & \end{bmatrix}$$

$$\frac{d\mathbf{A}(t)}{dt} = \begin{bmatrix} -2e^{-2t} & 3 \\ 2t & 0 \end{bmatrix}$$

Using the definition of the derivative of a matrix and the definition of addition we can show that

$$\frac{d}{dt}[\mathbf{A}(t) + \mathbf{B}(t)] = \frac{d\mathbf{A}(t)}{dt} + \frac{d\mathbf{B}(t)}{dt}$$

8-24

If t is a real variable and **A** is a matrix in which elements are constant, then

$$\frac{d}{dt}[t^k\mathbf{A}] = kt^{k-1}\mathbf{A}$$

If

$$\mathbf{B} = \begin{bmatrix} 3t^2 & 4t^2 \\ t^2 & 6t^2 \end{bmatrix}$$

find $\dfrac{d\mathbf{B}(t)}{dt}$

$$\frac{d\mathbf{B}(t)}{dt} = 2t \begin{bmatrix} 3 & 4 \\ 1 & 6 \end{bmatrix}$$

since

$$\mathbf{B} = t^2\mathbf{A} = t^2 \begin{bmatrix} 3 & 4 \\ 1 & 6 \end{bmatrix}$$

SUMMARY

1. The transformation of a vector in one set of coordinates into a vector in another set of coordinates is represented by

$$\mathbf{x} = \mathbf{T}\mathbf{y}$$

where **T** is a transformation matrix.

2. The n characteristic vectors of a matrix **A** can be arranged as the columns of a square matrix **M**, which is called the modal matrix.

$$\mathbf{M} = [\mathbf{x}_1, \mathbf{x}_2, \ldots, \mathbf{x}_n]$$

where $\mathbf{x}_i$ is the ith characteristic vector.

3. A matrix related to the characteristic roots of the matrix **A** is the spectral matrix **S** which is a diagonal matrix with the characteristic roots along the major diagonal so that

$$\mathbf{S} = \begin{bmatrix} \lambda_1 & 0 & \ldots & 0 \\ 0 & \lambda_2 & \ldots & 0 \\ \vdots & \vdots & \ddots & \vdots \\ 0 & 0 & & \lambda_n \end{bmatrix}$$

4. The characteristic roots of the spectral matrix and the characteristic roots of the original matrix **A** are the same.

5. The spectral matrix **S** is related to the modal matrix **M** as follows:

$$\mathbf{S} = \mathbf{M}^{-1}\mathbf{A}\mathbf{M}$$

where **A** is the original matrix.

6. The equation

$$\mathbf{S} = \mathbf{M}^{-1}\mathbf{AM}$$

is said to represent the diagonalization or transformation of the matrix **A** into a diagonal matrix.

7. A square matrix **A** of order n can be diagonalized if the matrix has n linearly independent characteristic vectors.

8. A matrix **A** always has n linearly independent vectors if it has n distinct characteristic roots or it is a symmetric matrix.

9. The two matrices **S** and **A** are said to be similar if there is a nonsingular matrix **M** such that

$$\mathbf{S} = \mathbf{M}^{-1}\mathbf{AM}$$

10. If **A** is a symmetric matrix with real elements, then the normalized modal matrix **N** associated with **A** is an orthogonal matrix and

$$\mathbf{N}' = \mathbf{N}^{-1}$$

11. A matrix polynomial written in terms of the matrix **A** is

$$P(\mathbf{A}) = c_0\mathbf{I} + c_1\mathbf{A} + c_2\mathbf{A}^2 + \ldots + c_n\mathbf{A}^n$$

where $\mathbf{I}$ = identity matrix and $\mathbf{A}^k$ implies that **A** is multiplied by itself k times. $P(\mathbf{A})$ is a matrix itself.

12. The Cayley-Hamilton theorem states that a matrix satisfies its own characteristic equation; that is

$$f(\mathbf{A}) = \mathbf{0}$$

where $f(\lambda) = c_0 + c_1\lambda + c_2\lambda^2 + \ldots + c_n\lambda^n$

13. An important polynomial function of **A**, the matrix exponential function is defined as

$$\exp(\mathbf{A}) = e^{\mathbf{A}} = \mathbf{I} + \frac{\mathbf{A}}{1!} + \frac{\mathbf{A}^2}{2!} + \ldots + \frac{\mathbf{A}^k}{k!} + \ldots$$

$$= \sum_{k=0}^{\infty} \frac{\mathbf{A}^k}{k!}$$

14. If **A** and **B** are commutative matrices, that is

$$\mathbf{AB} = \mathbf{BA}, \text{ then we have}$$

$$e^{\mathbf{A}} \cdot e^{\mathbf{B}} = e^{\mathbf{B}} \cdot e^{\mathbf{A}} = e^{(\mathbf{A}+\mathbf{B})}$$

15. It may be shown that exp(−**A**) is the inverse of the matrix

$$\exp(\mathbf{A}) \text{ or}$$

$$(\exp(\mathbf{A}))^{-1} \cdot \exp(\mathbf{A}) = \exp(-\mathbf{A}) \cdot \exp(\mathbf{A}) = \mathbf{I}$$

16. The derivative of a matrix $\mathbf{A}(t)$, where elements are functions of the variable t, is defined as

$$\frac{d}{dt}\mathbf{A}(t) = \begin{bmatrix} \dfrac{da_{11}(t)}{dt} & \ldots & \dfrac{da_{1n}(t)}{dt} \\ \cdot & & \cdot \\ \cdot & & \cdot \\ \cdot & & \cdot \\ \dfrac{da_{n1}(t)}{dt} & & \dfrac{da_{nn}(t)}{dt} \end{bmatrix}$$

EXERCISES

1. The vector $\mathbf{y} = \begin{bmatrix} 2 \\ 1 \end{bmatrix}$ is transformed into the z plane by the matrix

$$\mathbf{T} = \begin{bmatrix} 6 & 2 \\ 3 & 1 \end{bmatrix}$$

where $\mathbf{z} = \mathbf{T}\mathbf{y}$, find $\mathbf{z}$.

2. Consider the matrix

$$\mathbf{A} = \begin{bmatrix} 7 & -2 & 1 \\ -2 & 10 & -2 \\ 1 & -2 & 7 \end{bmatrix}$$

where the characteristic equation is

$$(\lambda - 6)^2 \ (\lambda - 12) = 0$$

and the characteristic vectors are

$$\mathbf{x}_1 = \begin{bmatrix} 1 \\ 0 \\ -1 \end{bmatrix}, \mathbf{x}_2 = \begin{bmatrix} 1 \\ 1 \\ 1 \end{bmatrix}, \mathbf{x}_3 = \begin{bmatrix} 1 \\ -2 \\ 1 \end{bmatrix}$$

The modal matrix is

$$\mathbf{M} =$$

3. Associated with matrix **A** in problem 2 is the spectral matrix

$$\mathbf{S} =$$

4. Obtain the normalized modal matrix **N** for matrix **A** in problem 2.

5. Given the modal matrix **M** for the matrix **A**, one may write an equation in terms of **M** and the spectral matrix as

$$\mathbf{S} =$$

6. The diagonal spectral matrix can always be obtained for a:

   lower triangular matrix
   symmetric matrix
   singular matrix

   Cross out the improper answers.

7. A square matrix **A** of order 4 can be diagonalized if the matrix has ________ linearly independent characteristic vectors.

8. If **A** is a symmetric matrix with real elements, then the normalized modal matrix **N** associated with **A** is a ________________________ matrix and one can write

$$\mathbf{N}^{-1} =$$

9. The Cayley-Hamilton theorem states that

10. Find the matrix exponential exp (**B**) when

$$\mathbf{B} = \begin{bmatrix} -t & 0 \\ 0 & 2t \end{bmatrix}$$

Hint: Recall that

$$e^{at} = 1 + at + \frac{(at)^2}{2!} + \ldots + \frac{(at)^k}{k!}$$

where *at* is a scalar variable.

11. Find the derivative of **Q**(t) when

$$\mathbf{Q}(t) = \begin{bmatrix} (3+t) & (6t^2+1) \\ e^{-2t} & 5 \end{bmatrix}$$

$$\frac{dQ(t)}{dt} =$$

See page 249 for the correct answers to the exercises.

A FINAL EXAMINATION

This final examination is a test of the knowledge and skill you have gained by completing this textbook. You should allow ninety minutes for all thirty questions.

1. A specific matrix **C** is written as

$$\mathbf{C} = \begin{bmatrix} 8 & 3 & -2 & 6 \\ 1 & 4 & 2 & -2 \\ 5 & 0 & -1 & 0 \end{bmatrix}$$

(a) The element c_{23} is equal to

(b) The matrix **C** is of order

2. Write a null matrix **N** of order 3 X 2.

$$\mathbf{N} = \begin{bmatrix} \quad & \quad \\ & \\ & \end{bmatrix}$$

3. Given

$$\mathbf{M} = \begin{bmatrix} -4j & (-1 + 3j) \\ 5 & +j \end{bmatrix}$$

Write the conjugate **M**.

4. The matrices **A**, **B**, **C**, **D**, **E**, and **F** are given below. Insert the letter identifying the correct matrix next to its proper name.

$$\mathbf{A} = \begin{bmatrix} 3 & 0 & 0 \\ 0 & 2 & 0 \end{bmatrix}$$

$$\mathbf{B} = \begin{bmatrix} 1 & 0 \\ 0 & 1 \end{bmatrix}$$

$$\mathbf{C} = [6 \quad 3 \quad 2]$$

$$\mathbf{D} = \begin{bmatrix} 4 & 2 & 1 \\ 0 & 6 & -1 \\ 0 & 0 & 3 \end{bmatrix}$$

$$\mathbf{E} = \begin{bmatrix} 3 \\ 4 \\ 1 \end{bmatrix}$$

$$\mathbf{F} = \begin{bmatrix} -1 & 0 & 0 \\ 0 & 3 & 0 \\ 0 & 0 & 2 \end{bmatrix}$$

Identity matrix ______________________

Upper triangular matrix ______________________

Diagonal matrix ______________________

Row vector ______________________

5. Given the matrix **Q**. Obtain **Q′** and tr(**Q**)

$$\mathbf{Q} = \begin{bmatrix} 6 & -1 & 3 \\ 2 & -2 & -4 \\ -3 & 1 & 5 \end{bmatrix}$$

$$\mathbf{Q}' = \begin{bmatrix} \quad & \quad & \quad \\ \quad & \quad & \quad \\ \quad & \quad & \quad \end{bmatrix}; \quad \mathrm{tr}(\mathbf{Q}) = \underline{\qquad\qquad}$$

6. Given two matrices **A** and **B**. Obtain **A + B** and **A – B**.

$$\mathbf{A} = \begin{bmatrix} 6 & 3 & 1 \\ 4 & -1 & 5 \end{bmatrix} ; \quad \mathbf{B} = \begin{bmatrix} 3 & 2 & 1 \\ -2 & 2 & 6 \end{bmatrix}$$

$$\mathbf{A} + \mathbf{B} = \begin{bmatrix} \quad & \quad & \quad \\ \quad & \quad & \quad \end{bmatrix} \qquad \mathbf{A} - \mathbf{B} = \begin{bmatrix} \quad & \quad & \quad \\ \quad & \quad & \quad \end{bmatrix}$$

7. The matrices **A** and **B** of problem 6 are conformable for multiplication as **AB**. True or False?

8. Complete the multiplication $\mathbf{yz}'$ when

$$\mathbf{y} = \begin{bmatrix} 3 \\ 2 \end{bmatrix} \quad \text{and} \quad \mathbf{z} = \begin{bmatrix} 4 \\ -2 \end{bmatrix}$$

9. For the matrices of problem 8,

$$\mathbf{yz}' = \mathbf{zy}' \qquad \text{True or False?}$$

10. A set of equations is

$$3x_1 + 6x_2 = 3$$

$$4x_1 + 9x_2 = 6$$

Solve for x_1 using Cramer's rule. Repeat using the elimination method.

11. Can you obtain the determinant of the matrix **B** when

$$\mathbf{B} = \begin{bmatrix} 6 & 4 & 1 \\ 3 & 2 & 0 \end{bmatrix}? \qquad \det \mathbf{B} = \underline{\hspace{4cm}}$$

12. Obtain the cofactor α_{21} of the matrix **P** where

$$\mathbf{P} = \begin{bmatrix} p_{11} & p_{12} & p_{13} \\ p_{21} & p_{22} & p_{23} \\ p_{31} & p_{32} & p_{33} \end{bmatrix}$$

13. Given two matrices **M** and **N**.

(a) det **M**′ = –det **M** True or False? ________________

(b) det **MN** = det **M** det **N** True or False? ________________

14. The rank of the matrix **Q** is ____________________ where

$$\mathbf{Q} = \begin{bmatrix} 8 & 1 & 0 & 0 \\ 2 & 2 & 2 & 8 \\ 37 & 2 & 5 & 14 \end{bmatrix}$$

15. A set of equations is represented by

$$\mathbf{Ax} = \mathbf{b}$$

where

$$\mathbf{A} = \begin{bmatrix} 4 & -1 \\ 8 & -2 \end{bmatrix}, \quad \mathbf{b} = \begin{bmatrix} 3 \\ 1 \end{bmatrix}$$

The matrix **A** is __________________ .
singular/nonsingular

The set of equation is ____________________ .
consistent/inconsistent

There is a __________________ solution.
unique/nonunique

16. Given the matrix **A**. Find the cofactor matrix $\mathbf{A}^c$ and the adjoint matrix adj **A**.

$$\mathbf{A} = \begin{bmatrix} 6 & 0 & 4 \\ 2 & 1 & 0 \\ 0 & -1 & 5 \end{bmatrix}$$

$$\mathbf{A}^c = \begin{bmatrix} \quad & \quad & \quad \end{bmatrix} \qquad \text{adj } \mathbf{A} = \begin{bmatrix} \quad & \quad & \quad \end{bmatrix}$$

17. Choose the proper answer to complete the relationship for the determinant of **A**.

$$|\mathbf{A}| \cdot |adj\ \mathbf{A}| =$$

$$\mathbf{A}^n,\ |\mathbf{A}|^{n-1},\ |\mathbf{A}|^n,\ \mathbf{A}^{n-1}$$

18. Obtain the inverse of the matrix **P** when

$$\mathbf{P} = \begin{bmatrix} -3 & 0 & 0 \\ 0 & 6 & 0 \\ 0 & 0 & 2 \end{bmatrix} \qquad \mathbf{P}^{-1} = \begin{bmatrix} \quad & \quad & \quad \end{bmatrix}$$

19. The definition of the inverse of a matrix **Q** is such that

____________________ $= \mathbf{I}$

20. Find the inverse of the matrix **A** in problem 16.

$$\mathbf{A}^{-1} = \begin{bmatrix} \quad & \quad & \quad \end{bmatrix}$$

21. Solve the set of simultaneous equations

$$3x_1 + 6x_2 = 2$$

$$4x_1 + 9x_2 = 6$$

by utilizing the matrix inversion operation.

22. Determine the solution of the equation $\mathbf{Ax} = \mathbf{0}$

when

$$\mathbf{A} = \begin{bmatrix} 6 & 3 \\ 2 & 4 \end{bmatrix}$$

23. The definition of an orthogonal matrix is such that given a matrix **P**

$$\mathbf{P} \cdot \underline{\qquad\qquad} = \mathbf{I}$$

24. The characteristic roots of an orthogonal matrix are equal to

________________________.

25. Determine the characteristic roots and vectors for the matrix

$$A = \begin{bmatrix} 6 & -6 \\ 4 & -4 \end{bmatrix}$$

Are the characteristic vectors linearly independent?

26. Determine the characteristic roots of

$$Q = \begin{bmatrix} 3 & 0 & 0 \\ 0 & -1 & 0 \\ 0 & 0 & -8 \end{bmatrix}$$

27. The vector $\mathbf{y} = -\begin{bmatrix} 3 \\ 1 \end{bmatrix}$

was obtained from the transformation $\mathbf{y} = \mathbf{Tx}$

where $\mathbf{T} = \begin{bmatrix} 2 & 1 \\ 1 & 1 \end{bmatrix}$

Determine **x**.

28. Consider a matrix **A** that has the characteristic equation

$$(\lambda - 5)\ (\lambda - 4)\ (\lambda - 3) = 0$$

and the characteristic vectors

$$\mathbf{x}_1 = \begin{bmatrix} 1 \\ 0 \\ 1 \end{bmatrix}, \mathbf{x}_2 = \begin{bmatrix} 1 \\ 1 \\ 1 \end{bmatrix}, \mathbf{x}_3 = \begin{bmatrix} 1 \\ -3 \\ 1 \end{bmatrix}$$

The modal matrix is

$$\mathbf{M} = \begin{bmatrix} \quad \end{bmatrix}$$

The normalized modal matrix is

$$\mathbf{N} = \begin{bmatrix} \quad \end{bmatrix}$$

The associated spectral matrix is

$$\mathbf{S} = \begin{bmatrix} \quad \end{bmatrix}$$

29. The diagonal spectral matrix can always be obtained for a ____________ matrix.

30. A square matrix of order m can be diagonalized if the matrix has ____________ linearly independent characteristic vectors.

The answers to the final examination appear on page 251.

ANSWERS TO EXERCISES

Chapter 1

1. **C** is not a matrix because the elements are not within the proper bracket or parenthesis symbols. Also, the array is not rectangular because the element c_{21} is missing.

2. 3 X 2

3. **A**

4. 0

5.
$$\mathbf{M} = \begin{bmatrix} 41 & 73 & 172 \\ 33 & 68 & 158 \end{bmatrix}$$

Matrix **M** is of order 2 X 3.

If you correctly answered three or more questions you are ready to proceed to the next chapter on page 21. If you answered less than three questions correctly, you probably should review this section.

Chapter 2

1. **A** = 4**I**, a scalar matrix (p. 31)
B = a row vector (p. 21)
C = an identity matrix (p. 25)
D = a symmetric matrix (p. 29)
E = an upper triangular matrix (p. 26)

2. $\mathbf{0} = \begin{bmatrix} 0 & 0 & 0 \\ 0 & 0 & 0 \end{bmatrix}$ (p. 23)

3. $\overline{\mathbf{H}} = \begin{bmatrix} 7-j & -4j \\ j & 5 \end{bmatrix}$ (p. 32)

If you obtained five of the seven answers, you are ready to proceed to Chapter 3 on page 37. If not, it would be best to review the chapter, and pay attention to pages noted for each concept.

Chapter 3

1. The associative law of addition. (p. 43)

2. $\mathbf{xy}' = \begin{bmatrix} 10 & 4 \\ 5 & 2 \end{bmatrix}$ (p. 49)

3. $\mathbf{A}+\mathbf{B} = \begin{bmatrix} 12 & 1 \\ 2 & 2 \end{bmatrix}$ $\mathbf{A}-\mathbf{B} = \begin{bmatrix} 4 & 5 \\ 2 & 0 \end{bmatrix}$ (p. 41)

4. $\mathbf{AB} = \begin{bmatrix} 13 & 14 \\ 7 & 2 \end{bmatrix}$ (p. 48)

5. No. In general matrix multiplication is not commutative. (p. 55)

If you obtained 4 of the 5 answers, proceed to Chapter 4 on page 62. Otherwise, you should review the summary on page 58 and then try the following two exercises before proceeding to Chapter 4.

ADDITIONAL EXERCISES

6. Determine the product **MN′** where

$$\mathbf{M} = \begin{bmatrix} 3 & 2 \\ 1 & 0 \end{bmatrix}, \mathbf{N} = [1 \quad 2]$$

7. If **AB** = **C** and **AD** = **C** can we state that in general **B** = **D**?

6. $MN' = \begin{bmatrix} 3 & 2 \\ 1 & 0 \end{bmatrix} \begin{bmatrix} 1 \\ 2 \end{bmatrix} = \begin{bmatrix} 7 \\ 1 \end{bmatrix}$

7. If **AB** = **C** and **AD** = **C** we cannot state that **B** = **D** in general. (See page 58)

Now proceed to Chapter 4 on page 62.

Chapter 4

1. One cannot obtain the det **A** when **A** is not a square matrix. (p. 68)

2. $$\begin{bmatrix} 4 & 6 & 0 \\ 5 & 0 & 7 \\ 0 & 2 & 4 \end{bmatrix} \begin{bmatrix} x_1 \\ x_2 \\ x_3 \end{bmatrix} = \begin{bmatrix} 7 \\ 8 \\ 0 \end{bmatrix}$$ (p. 64)

3. $|\mathbf{M}_{22}| = \begin{bmatrix} p_{11} & p_{13} \\ p_{31} & p_{33} \end{bmatrix}$, $|\mathbf{M}_{22}|$ is a *second* order minor (p. 72)

4. det **R** = –6

$$\det \mathbf{R} = \begin{vmatrix} 5 & 0 & 2 \\ 1 & 6 & 2 \\ 3 & 0 & 1 \end{vmatrix} = 5 \begin{vmatrix} 6 & 2 \\ 0 & 1 \end{vmatrix} + 2(-1)^4 \begin{vmatrix} 1 & 6 \\ 3 & 0 \end{vmatrix}$$

$$= 5(6) + 2(-18)$$

$$= -6$$

by expansion of minors along the first row.

Using the procedure on page 71

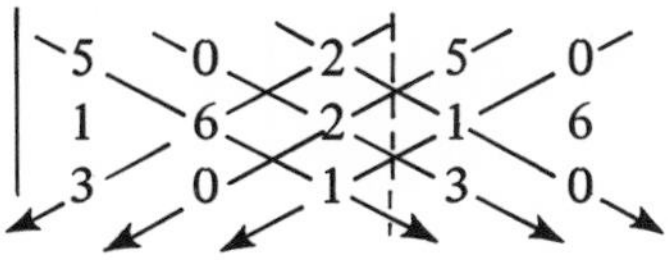

$$\det \mathbf{R} = 5(6)1 + 0(2)3 + 2(1)0 - 2(6)3 - 5(2)0 - 0(1)1$$
$$= 30 - 36 = -6$$

5. $\det \mathbf{Q} = 46$

$$\det \mathbf{Q} = \begin{vmatrix} 2 & 1 & 0 & 5 \\ 3 & 1 & 5 & 0 \\ 2 & 0 & 1 & 0 \\ 8 & 0 & 2 & 2 \end{vmatrix} = a_{31}\alpha_{31} + a_{33}\alpha_{33}$$

when we expand along the third row, where $a_{32} = a_{34} = 0$

$$\det \mathbf{Q} = 2(-1)^4 \begin{vmatrix} 1 & 0 & 5 \\ 1 & 5 & 0 \\ 0 & 2 & 2 \end{vmatrix} + 1(-1)^6 \begin{vmatrix} 2 & 1 & 5 \\ 3 & 1 & 0 \\ 8 & 0 & 2 \end{vmatrix}$$

$$= 2(\overbrace{1(10) + 5(2)}^{\text{row 1 of } \mathbf{M}_{31}}) + 1(\overbrace{8(-5) + 2(-1)}^{\text{row 3 of } \mathbf{M}_{33}})$$

$$= 2(20) + 1(-42)$$

$$= -2$$

(p. 74)

6. $x_3 = \frac{13}{7}$

$$\begin{bmatrix} 2 & 6 & 0 \\ 4 & 0 & 2 \\ 0 & 1 & 1 \end{bmatrix} \begin{bmatrix} x_1 \\ x_2 \\ x_3 \end{bmatrix} = \begin{bmatrix} 0 \\ 2 \\ 2 \end{bmatrix}$$

$$\mathbf{Ax} = \mathbf{b}$$

$$x_3 = \frac{\det \mathbf{A}_3}{\det \mathbf{A}}$$

$$\det \mathbf{A} = \begin{vmatrix} 2 & 6 & 0 \\ 4 & 0 & 2 \\ 0 & 1 & 1 \end{vmatrix} = 2(-2) + 6(-1)^3(4) = -28$$

$$\det \mathbf{A}_3 = \begin{vmatrix} 2 & 6 & 0 \\ 4 & 0 & 2 \\ 0 & 1 & 2 \end{vmatrix} = 2(-2) + 6(-1)8 = -52$$

$$x_3 = \frac{\det \mathbf{A}_3}{\det \mathbf{A}} = \frac{-52}{-28} = \frac{13}{7}$$

(p. 82)

7. $\det \mathbf{P}' = \det \mathbf{P}$ (p. 81)

8. $\det \mathbf{PQ} = \det \mathbf{P} \det \mathbf{Q}$ (p. 81)

9. The $\det \mathbf{A} = 0$ because the elements of the third row are proportionately related by a factor of 3 to the corresponding elements of the first row. (p. 81)

10. The matrix equation $\mathbf{Qy} = \mathbf{c}$ is called a nonhomogeneous equation. (p. 67)

11. $x_1 = 1, x_2 = 2$, and $x_3 = -1$.

The outline of the procedure is

Step 1. $x_1 = \frac{\overline{1}}{2}x_2 + \frac{\overline{1}}{2}x_3 + \frac{1}{2}$ (1)

Step 2.
$$\begin{cases} \frac{3}{2}x_2 + \frac{3}{2}x_3 + \frac{3}{2} + x_2 + x_3 = 4 & (2) \\ x_2 + x_3 + 1 + 2x_2 - 2x_3 = 8 & (3) \end{cases}$$

Step 3. Solving equation (2) for x_2,

$$x_2 = -x_3 + 1 \quad (4)$$

Step 4. Substituting equation (4) into equation (3)

$3(-x_3 + 1) - x_3 = 7$ or $x_3 = -1$

Step 5. Substituting $x_3 = -1$ into equation (4),

$$x_2 = -(-1) + 1 = 2$$

Step 6. Substituting x_2 and x_3 into equation (1), we obtain

$$x_1 = 1$$

Proceed to Chapter 5. on page 96.

Chapter 5

1. $r(\mathbf{P}) = 2$

since

$$\begin{vmatrix} 1 & 2 \\ -4 & 0 \end{vmatrix} = 8$$

and there are no minors of order 3. (p. 96)

2. The matrix **A** is singular since det **A** = 0. (p. 98)

The set of equations is inconsistent since $r(\mathbf{A}) = 2$ and the $r(\mathbf{A}^b) = 3$.

$$\det \mathbf{A} = 0 \quad \text{and} \quad \begin{vmatrix} 2 & 3 \\ 2 & 5 \end{vmatrix} \neq 0 \quad \text{so} \quad r(\mathbf{A}) = 2$$

$$\mathbf{A}^b = \begin{bmatrix} 1 & 2 & 3 & 1 \\ 1 & 2 & 5 & 3 \\ 2 & 4 & 8 & 0 \end{bmatrix}, \quad \begin{vmatrix} 1 & 3 & 1 \\ 1 & 5 & 3 \\ 2 & 8 & 0 \end{vmatrix} = -8$$

so $r(\mathbf{A}^b) = 3$

3. There is a unique solution since

$$r(\mathbf{A}) = 2 \quad \text{and} \quad r(\mathbf{A}^b) = 2$$ (p. 103)

4. tr (**A**) =8 (p. 107)

5.

$$\mathbf{A}^c = \begin{bmatrix} 2 & -2 & -4 \\ -10 & 0 & 5 \\ -14 & 4 & 8 \end{bmatrix}$$ (p. 108)

6. $\text{adj } \mathbf{A} = \mathbf{A}^{c'} = \begin{bmatrix} 2 & -10 & -14 \\ -2 & 0 & 4 \\ -4 & 5 & 8 \end{bmatrix}$ (p. 111)

7. $\mathbf{A} \cdot (\text{adj } \mathbf{A}) = |\mathbf{A}| \, \mathbf{I}$ (p. 113)

8. $|\mathbf{A}| \cdot |\text{adj } \mathbf{A}| = |\mathbf{A}|^n$ (p. 115)

You should have correctly answered 6 or more of the questions. If you did, proceed to Chapter 6 on page 121. If you answered less than six correctly, reread the review of Chapter 5 on pages 117-118 and the specific pages noted after each question you missed; then try the exercise questions again before proceeding.

Chapter 6

1. $(\mathbf{DP})^{-1} = \mathbf{P}^{-1}\mathbf{D}^{-1}$ (p. 134)

2.

$$\mathbf{Q}^{-1} = \frac{1}{23(6)}\begin{bmatrix} -2 & 3 & 1 \\ -6 & 0 & 6 \\ 4 & 0 & -2 \end{bmatrix}$$ (p. 133)

where

$$\mathbf{Q} = 23\begin{bmatrix} 0 & 1 & 3 \\ 2 & 0 & 1 \\ 0 & 2 & 3 \end{bmatrix} = 23\mathbf{A}$$

and det $\mathbf{A} = 6$

3.

$$\mathbf{P}^{-1} = \begin{bmatrix} \frac{1}{18} & 0 & 0 & 0 \\ 0 & \frac{1}{21} & 0 & 0 \\ 0 & 0 & \frac{1}{13} & 0 \\ 0 & 0 & 0 & \frac{1}{5} \end{bmatrix}$$ (p. 132)

4. $\mathbf{D}^{-1}\mathbf{D} = \mathbf{D}\mathbf{D}^{-1} = \mathbf{I}$ (p. 121)

5.

$$\mathbf{A}^{-1} = \begin{bmatrix} \frac{1}{7} & \frac{-1}{14} & 0 \\ \frac{-3}{98} & \frac{31}{196} & \frac{-1}{14} \\ \frac{-1}{98} & \frac{-3}{98} & \frac{1}{7} \end{bmatrix}$$

since

$$\mathbf{A}^c = \begin{bmatrix} 56 & -12 & -4 \\ -28 & 62 & -12 \\ 0 & -28 & 56 \end{bmatrix} \quad \text{and det } \mathbf{A} = 392$$ (p. 126)

6.

$$\mathbf{x} = \mathbf{A}^{-1}\mathbf{b} = \begin{bmatrix} 0 \\ 27 \\ -5 \end{bmatrix}$$

since

$$\mathbf{x} = \mathbf{A}^{-1}\mathbf{b} = \begin{bmatrix} \frac{1}{7} & \frac{-1}{14} & 0 \\ \frac{-3}{98} & \frac{31}{196} & \frac{-1}{14} \\ \frac{-1}{98} & \frac{-3}{98} & \frac{1}{7} \end{bmatrix} \begin{bmatrix} 98 \\ 196 \\ 14 \end{bmatrix} = \begin{bmatrix} (14 - 14 + 0) \\ (-3 + 31 - 1) \\ (-1 - 6 + 2) \end{bmatrix}$$

where $\mathbf{A}^{-1}$ was obtained in problem 5. (p. 138)

7. False. If $\mathbf{P}$ is a nonsingular matrix and

$$\mathbf{PA} = \mathbf{BP}$$

then the statement $\mathbf{A} = \mathbf{B}$ is false.

Note that

$$\mathbf{PA} = \mathbf{BP}$$

premultiplied by $\mathbf{P}^{-1}$ yields

$$\mathbf{P}^{-1}\mathbf{PA} = \mathbf{P}^{-1}\mathbf{BP}$$

or

$$\mathbf{A} = \mathbf{P}^{-1}\mathbf{BP}$$

8. The solution of $\mathbf{Ax} = \mathbf{0}$ is $\mathbf{x} = \mathbf{0}$

when

$$\mathbf{A} = \begin{bmatrix} 1 & 0 & 1 \\ 0 & 1 & -2 \\ 1 & 0 & -1 \end{bmatrix}$$

since

$$\det \mathbf{A} = 1(-1) + 1(-1) = -2$$

and $\mathbf{A}$ is a nonsingular square matrix. (p. 140)

9. The solution of $\mathbf{Ax} = \mathbf{0}$

is

$$\mathbf{x} = \beta \begin{bmatrix} -1 \\ 1 \\ 1 \end{bmatrix}$$

when

$$\mathbf{A} = \begin{bmatrix} 1 & 0 & 1 \\ 0 & 1 & -1 \\ 1 & 0 & 1 \end{bmatrix}$$

since

$$\det \mathbf{A} = 1(1) + 1\,(-1) = 0$$

and $\mathbf{A}$ is a singular square matrix.

Solving the first row we have

$$x_1 + x_3 = 0 \quad \text{or} \quad x_1 = -x_3$$

Let $x_3 = \beta$, an arbitrary constant.

Then

$$x_1 = -\beta$$

and from the second row we have

$$x_2 - x_3 = 0 \text{ or } x_2 = x_3 = \beta$$ (p. 141)

10. $\mathbf{QQ}' = \mathbf{I}$ and therefore $\mathbf{Q}$ is an orthogonal matrix

$$\begin{bmatrix} \frac{1}{\sqrt{3}} & \frac{1}{\sqrt{6}} & \frac{1}{\sqrt{2}} \\ \frac{1}{\sqrt{3}} & \frac{-2}{\sqrt{6}} & 0 \\ \frac{1}{\sqrt{3}} & \frac{1}{\sqrt{6}} & \frac{1}{\sqrt{2}} \end{bmatrix} \begin{bmatrix} \frac{1}{\sqrt{3}} & \frac{1}{\sqrt{3}} & \frac{1}{\sqrt{3}} \\ \frac{1}{\sqrt{6}} & \frac{-2}{\sqrt{6}} & \frac{1}{\sqrt{6}} \\ \frac{-1}{\sqrt{2}} & 0 & \frac{1}{\sqrt{2}} \end{bmatrix} = \begin{bmatrix} 1 & 0 & 0 \\ 0 & 1 & 0 \\ 0 & 0 & 1 \end{bmatrix}$$

(p. 137)

Proceed to Chapter 7 on page 149.

Chapter 7

1. $\lambda_1 = 12, \lambda_2 = 24, \lambda_3 = 36$

(p. 178)

2. $\lambda_1 = 1, \lambda_2 = \frac{1}{2}, \lambda_3 = \frac{1}{3}$

(p. 178)

3. The characteristic roots of an orthogonal matrix are equal to either +1 or −1.

(p. 180)

4. The sum of the characteristic roots of **B** is equal to tr(B) = 9 (p. 185)

5. $\lambda_1 = 4, \lambda_2 = -3, \lambda_3 = 2$ (p. 182)

6.

$$\det \mathbf{M} = \begin{vmatrix} 4 & 2 & 6 \\ 2 & 1 & 3 \\ 0 & -3 & -5 \end{vmatrix} = 4(4) - 2(8) = 0$$

Since the rank of **M** is less than 3, the vectors are dependent. Row 1 and row 2 are dependent.

(p. 166)

7. $\det \mathbf{Q} = 3$

$$\det(\mathbf{Q} - \lambda\mathbf{I}) = \det \begin{bmatrix} (2-\lambda) & 1 & 1 \\ 1 & (2-\lambda) & 1 \\ 0 & 0 & (1-\lambda) \end{bmatrix} = (1-\lambda)\left((2-\lambda)^2 - 1\right)$$

$$= \lambda^3 - 5\lambda^2 + 7\lambda - 3 = 0$$

The roots are

$$\lambda_1 = 1, \lambda_2 = 1, \lambda_3 = 3$$

For $\lambda_1 = \lambda_2 = 1$, we have

$$(\mathbf{Q} - \mathbf{I})\mathbf{x} = \mathbf{0}$$

or

$$\begin{bmatrix} 1 & 1 & 1 \\ 1 & 1 & 1 \\ 0 & 0 & 0 \end{bmatrix} \mathbf{x} = \mathbf{0}, \quad \det(\mathbf{Q} - \mathbf{I}) = 0,$$

so set one element equal to zero in each vector, obtaining

$$\mathbf{x}_1 = \beta \begin{bmatrix} 1 \\ 0 \\ -1 \end{bmatrix} \text{ and } \mathbf{x}_2 = \gamma \begin{bmatrix} 0 \\ 1 \\ -1 \end{bmatrix}$$

Note that $\mathbf{x}_1$ and $\mathbf{x}_2$ are linearly independent.

For $\lambda_3 = 3$, we obtain

$$(\mathbf{Q} - 3\mathbf{I})\mathbf{x}_3 = \mathbf{0}, \text{ or } \begin{bmatrix} -1 & 1 & 1 \\ 1 & -1 & 1 \\ 0 & 0 & -2 \end{bmatrix} \mathbf{x}_3 = \mathbf{0}$$

The last row implies that $\mathbf{x}_3 = 0$ and thus $x_1 = x_2$. Therefore,

$$\mathbf{x}_3 = \delta \begin{bmatrix} 1 \\ 1 \\ 0 \end{bmatrix}$$

The vectors $\mathbf{x}_1, \mathbf{x}_2$, and $\mathbf{x}_3$ are linearly independent since $\mathbf{M} = [x_1, x_2, x_3]$ is of rank 3.

(p. 165)

8.

$$\mathbf{x}_1 = \beta \begin{bmatrix} -4 \\ 1 \\ 0 \end{bmatrix}$$

$$(\mathbf{N} - \lambda_1\mathbf{I})\mathbf{x}_1 = \mathbf{N}\mathbf{x}_1 = \mathbf{0}$$

since $\lambda_1 = 0$. The determinant of $\mathbf{N}$ is

$$\det \mathbf{N} = 1 \begin{vmatrix} -2 & -8 \\ 1 & 4 \end{vmatrix} = 0$$

Since $\mathbf{N}$ is singular, a nonzero vector can be obtained.

$$\mathbf{N}\mathbf{x} = \begin{bmatrix} -2 & -8 & -12 \\ 1 & 4 & 4 \\ 0 & 0 & 1 \end{bmatrix} \mathbf{x} = \mathbf{0}$$

The third row yields $x_3 = 0$. The first or second row yields

$$-2x_1 - 8x_2 = 0, \text{ or } x_1 = -4x_2$$

Therefore,

$$\mathbf{x}_1 = \beta \begin{bmatrix} -4 \\ 1 \\ 0 \end{bmatrix}$$

(p. 172)

9. The characteristic roots of a real symmetric matrix are all real. (p. 161)

Proceed to Chapter 8, page 186.

Chapter 8

1.

$$\mathbf{z} = \begin{bmatrix} 14 \\ 7 \end{bmatrix}$$

since

$$\mathbf{z} = \mathbf{Ty} = \begin{bmatrix} 6 & 2 \\ 3 & 1 \end{bmatrix} \begin{bmatrix} 2 \\ 1 \end{bmatrix} = \begin{bmatrix} 14 \\ 7 \end{bmatrix}$$

(p. 190)

2.

$$\mathbf{M} = \begin{bmatrix} 1 & 1 & 1 \\ 0 & 1 & -2 \\ -1 & 1 & 1 \end{bmatrix}$$

(p. 195)

3.

$$\mathbf{S} = \begin{bmatrix} 6 & 0 & 0 \\ 0 & 6 & 0 \\ 0 & 0 & 12 \end{bmatrix}$$

(p. 196)

4.

$$\mathbf{N} = \begin{bmatrix} \frac{1}{\sqrt{2}} & \frac{1}{\sqrt{3}} & \frac{1}{\sqrt{6}} \\ 0 & \frac{1}{\sqrt{3}} & \frac{-2}{\sqrt{6}} \\ \frac{1}{\sqrt{2}} & \frac{1}{\sqrt{3}} & \frac{1}{\sqrt{6}} \end{bmatrix}$$

(p. 206)

5. $\mathbf{S} = \mathbf{M}^{-1}\mathbf{AM}$ (p. 201)

6. The diagonal spectral matrix can always be obtained for a symmetric matrix. (p. 204)

7. A square matrix of order 4 can be diagonalized if the matrix has 4 linearly independent characteristic vectors. (p. 201)

8. If **A** is a symmetric matrix with real elements, then the normalized modal matrix **N** associated with **A** is an orthogonal matrix and one can write

$$\mathbf{N}^{-1} = \mathbf{N}'$$ (p. 206)

9. The Cayley-Hamilton theorem states that a matrix satisfies its own characteristic equation; that is

$$f(\mathbf{A}) = \mathbf{0}$$

when $f(\lambda) = 0$ is the characteristic equation. (p. 210)

10.

$$\exp(\mathbf{B}) = \begin{bmatrix} e^{-t} & 0 \\ 0 & e^{2t} \end{bmatrix}$$

recalling that

$$e^{at} = 1 + at + \frac{(at)^2}{2!} + \ldots + \frac{(at)^k}{k!}$$

which is the exponential expansion in the scalar variable at.

Then

$$\exp(\mathbf{B}) = \mathbf{I} + \mathbf{B} + \frac{\mathbf{B}^2}{2!} + \ldots =$$

$$\begin{bmatrix} 1 & 0 \\ 0 & 1 \end{bmatrix} + \begin{bmatrix} -t & 0 \\ 0 & +2t \end{bmatrix} + \frac{1}{2}\begin{bmatrix} t^2 & 0 \\ 0 & 4t^2 \end{bmatrix} + \ldots = \begin{bmatrix} e^{t} & 0 \\ 0 & e^{2t} \end{bmatrix}$$

(p. 212)

11.

$$\frac{d\,\mathbf{Q}(t)}{dt} = \begin{bmatrix} 1 & 12t \\ -2e^{-2t} & 0 \end{bmatrix}$$

(p. 218)

Congratulations! You have completed the eight chapters of this text. In order to test your knowledge of the algebra of matrices proceed to the final examination on page 227.

ANSWERS TO THE FINAL EXAMINATION

1. 2 points

(a) The element c_{23} is equal to +2. (p. 11)
(b) The matrix **C** is of order 3 X 4. (p. 12)

2. 2 points

$$\mathbf{N} = \begin{bmatrix} 0 & 0 \\ 0 & 0 \\ 0 & 0 \end{bmatrix}$$

(p. 23)

3. 2 points

$$\overline{\mathbf{M}} = \begin{bmatrix} 4j & (-1-3j) \\ 5 & -j \end{bmatrix}$$

(p. 32)

4. 4 points

Identity matrix **B** (p. 25)
Upper triangular matrix **D** (p. 26)
Diagonal matrix **F** (p. 24)
Row vector **C** (p. 21)

5. 4 points

$$Q' = \begin{bmatrix} 6 & 2 & -3 \\ -1 & -2 & 1 \\ 3 & -4 & 5 \end{bmatrix}$$

$\text{tr}\,(\mathbf{Q}) = 9$ (p. 107)

6. 4 points

$$\mathbf{A} + \mathbf{B} = \begin{bmatrix} 9 & 5 & 2 \\ 2 & 1 & 11 \end{bmatrix}$$

(p. 41)

$$\mathbf{A} - \mathbf{B} = \begin{bmatrix} 3 & 1 & 0 \\ 6 & -3 & -1 \end{bmatrix}$$

(p. 45)

7. 2 points

False. The number of rows of **B** is not equal to the number of columns of **A**.
(p. 52)

8. 2 points

$$\mathbf{yz}' = \begin{bmatrix} 12 & -6 \\ 8 & -4 \end{bmatrix}$$

(p. 54)

9. 2 points

False

(p. 54)

10. 3 points

$$x_1 = -3$$

By Cramers' rule, since

$$\begin{bmatrix} 3 & 6 \\ 4 & 9 \end{bmatrix} \mathbf{x} = \begin{bmatrix} 3 \\ 6 \end{bmatrix}$$

$$x_1 = \frac{\det \mathbf{A}_1}{\det \mathbf{A}} = \frac{-9}{3}$$ (p. 82)

By the elimination method:

(1) $x_1 = -\ 2x_2 + 1$

(2) $4(-2x_2 + 1) + 9x_2 = 6$

(3) $x_2 = 2$

(4) $x_1 = 3$ (p. 86)

11. 2 points

No. The determinant can only be obtained for square matrices. (p. 68)

12. 3 points

$$a_{21} = (-1)^3 \begin{vmatrix} p_{12} & p_{13} \\ p_{32} & p_{33} \end{vmatrix} = -(p_{12}p_{33} - p_{13}p_{32})$$

(p. 73)

13. 2 points

(a) False. $\det \mathbf{M}' = \det \mathbf{M}$ (p. 81)
(b) True (p. 81)

14. 2 points

The rank of the matrix **Q** is 3. (p. 96)

15. 3 points

The matrix **A** is *singular*. (p. 98)
The set of equations is *inconsistent*. (p. 101)
There is a *nonunique* solution. (p. 103)

16. 4 points

$$\mathbf{A}^c = \begin{bmatrix} 5 & -10 & -2 \\ -4 & 30 & 6 \\ -4 & 8 & 6 \end{bmatrix}$$ (p. 108)

$$\text{adj } \mathbf{A} = \begin{bmatrix} 5 & -4 & -4 \\ -10 & 30 & 8 \\ -2 & 6 & 6 \end{bmatrix}$$ (p. 111)

17. 2 points

$$|\mathbf{A}| \cdot |\text{adj } \mathbf{A}| = |\mathbf{A}|^n$$ (p. 115)

18. 2 points

$$\mathbf{P}^{-1} = \begin{bmatrix} \frac{-1}{3} & 0 & 0 \\ 0 & \frac{1}{6} & 0 \\ 0 & 0 & \frac{1}{2} \end{bmatrix}$$ (p. 132)

19. 2 points

$$\mathbf{Q}^{-1}\mathbf{Q} = \mathbf{I}$$

or

$$\mathbf{Q}\mathbf{Q}^{-1} = 1$$ (p. 121)

20. 2 points

$$A^{-1} = \frac{1}{22}\begin{bmatrix} 5 & -4 & -4 \\ -10 & 30 & 8 \\ -2 & 6 & 6 \end{bmatrix}$$

since

$$\det \mathbf{A} = 22$$ (p. 133)

21. 3 points

$$\mathbf{x} = \mathbf{A}^{-1}\mathbf{b}$$

$$\mathbf{A}^{-1} = \frac{1}{3}\begin{bmatrix} 9 & -6 \\ -4 & 3 \end{bmatrix}, \mathbf{x} = \begin{bmatrix} -6 \\ \frac{10}{3} \end{bmatrix}$$ (p. 138)

22. 2 points

$\mathbf{x} = \mathbf{0}$ since det $\mathbf{A} \neq 0$. (p. 140)

23. 2 points

$$\mathbf{P} \cdot \mathbf{P}' = \mathbf{I}$$ (p. 136)

24. 2 points

The characteristic roots of an orthogonal matrix are equal to either +1 or –1. (p. 179)

25. 4 points

$$\det(\mathbf{A} - \lambda\mathbf{I}) = \begin{vmatrix} (6-\lambda) & -6 \\ 4 & (-4-\lambda) \end{vmatrix} = \lambda^2 - 2\lambda = \lambda(\lambda - 2) = 0$$

det **A** = 0

for

$$\lambda_1 = 0,\ x_1 = x_2, \text{ and } x_1 = \beta \begin{bmatrix} 1 \\ 1 \end{bmatrix}$$

For

$$\lambda_2 = 2,\ x_1 = \frac{3}{2} x_2 \text{ and } x_2 = \gamma \begin{bmatrix} \frac{3}{2} \\ 1 \end{bmatrix}$$

$$\det \mathbf{M} = \begin{bmatrix} 1 & \frac{3}{2} \\ 1 & 1 \end{bmatrix} \neq 0$$

and the vectors are linearly independent. (p. 166)

26. 3 points

$$\lambda_1 = 3,\ \lambda_2 = -1,\ \lambda_3 = -8$$

(p. 150)

27. 2 points

$$\mathbf{x} = \mathbf{T}^{-1}\mathbf{y} = \begin{bmatrix} 1 & -1 \\ -1 & 2 \end{bmatrix} \begin{bmatrix} 3 \\ 1 \end{bmatrix} = \begin{bmatrix} 2 \\ -1 \end{bmatrix}$$

(p. 193)

28. 3 points

$$M = \begin{bmatrix} 1 & 1 & 1 \\ 0 & 1 & -3 \\ 1 & 1 & 1 \end{bmatrix}$$

$$N = \begin{bmatrix} \frac{1}{\sqrt{2}} & \frac{1}{\sqrt{3}} & \frac{1}{\sqrt{11}} \\ 0 & \frac{1}{\sqrt{3}} & \frac{-3}{\sqrt{11}} \\ \frac{1}{\sqrt{2}} & \frac{1}{\sqrt{3}} & \frac{1}{\sqrt{11}} \end{bmatrix}$$

$$S = \begin{bmatrix} 5 & 0 & 0 \\ 0 & 4 & 0 \\ 0 & 0 & 3 \end{bmatrix}$$

(p. 196)

29. 2 points

The diagonal spectral matrix can always be obtained for a symmetric matrix. (p. 204)

30. 2 points

A square matrix of order m can always be diagonalized if the matrix has m linearly independent characteristic vectors. (p. 201)

Your Total Score

There is a total of 76 points obtainable in the examination. If you scored 68 points or more you have answered 90% or more of the questions correctly.

index

Addition of matrices, 41
Adjoint matrix, 111
Array, 3
Associative law of addition, 43
Augmented matrix, 101

Cancellation law for addition, 44
Cayley-Hamilton Theorem, 210
Characteristic,
 equation, 150
 polynomial, 150, 210
 root, 151
 value problem, 148
 vector, 155
Cofactor, 73
Cofactor matrix, 108
Column, 6
Column vector, 22
Commutative law of addition, 43
Commutative multiplication, 55
Conformable,
 for addition, 41
 for multiplication, 48
Conjugate of a matrix, 32
Consistent linear equations, 101
Cramer's rule, 82

Derivative of a matrix, 218
Determinant, 68
Diagonal matrix, 24
Diagonalization of a matrix, 201

Eigenvalue, 148, 150
Elements, 3, 11
Elimination method, 86
Equal matrices, 37
Expansion by cofactors, 74
Exponential function, 213

Function of a matrix, 210

Gaussian elimination method, 86

Homogeneous equation, 67, 139

Identity matrix, 25
Inconsistent linear equations, 101
Inverse of a matrix, 121

Linear algebraic equations, 62
Linear dependence, 165
Lower triangular matrix, 26

Matrix, 10
Matrix polynomial, 208
Minor, 72
Modal matrix, 195
Multiplication of matrices, 48

Newton-Raphson method, 153

Nonhomogeneous equation, 67
Nonsingular matrix, 98
Normalized modal matrix, 206
Null matrix, 23

Order, 12
Orthogonal matrix, 136, 179

Quantitative, 1

Rank of a matrix, 96
Row vector, 21

Scalar matrix, 31
Scalar quantity, 1
Similar matrices, 203
Singular matrix, 98
Spectral matrix, 196
Square matrix, 13
Subtraction of matrices, 45
Symmetric matrix, 29

Trace, 107
Transformation matrix, 190
Transpose of a matrix, 38
Triangular matrix, 26
Trivial solution, 140

Unique solution, 103
Upper triangular matrix, 26, 40

Vector, 21

Zero matrix, 23